ROTARY DRILLING SERIES

The Bit

Unit I, Lesson 2
Fourth Edition

By Kate Van Dyke

Published by

PETROLEUM EXTENSION SERVICE
Continuing & Extended Education
The University of Texas at Austin
Austin, Texas

in cooperation with

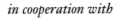

INTERNATIONAL ASSOCIATION
OF DRILLING CONTRACTORS
Houston, Texas

1995

Library of Congress Cataloging-in-Publication Data

Van Dyke, Kate, 1951–
    The bit / by Kate Van Dyke. — 4th ed.
        p.   cm. — (Rotary drilling ; unit 1, lesson 2)
    "In cooperation with the International Association of Drilling
Contractors, Houston, Texas."
    ISBN 0-88698-167-0
    1. Bits (Drilling and boring) 2. Oil well drilling—Equipment and
supplies. I. University of Texas at Austin. Petroleum Extension Service.
II. International Association of Drilling Contractors.   III. Title.
IV. Series: Rotary drilling series: unit 1, lesson 2.
    TN871.3.V269   1995
    622' .3381—dc20   95-9723
                                                                                        CIP

First Edition published 1966. Fourth Edition 1995
Third Impression 2000
Printed in the United States of America

Catalog no. 2.10240
ISBN 0-88698-167-0

Contents

▼
▼
▼

Figures   V

Foreword   IX

Acknowledgments   XI

Units of Measurement   XII

Introduction   1

ROP, Longevity, and Gauge   4

Formations and Bit Design   5

Choosing a Bit   7
  To Summarize   8

Roller Cone Bits   9
  History   12
  Types of Roller Cone Bits   13
    Tungsten Carbide Bits   13
  Roller Cone Bit Manufacture and Design   15
  Cones   16
    Cone Alignment   17
    Interfit   19
    Journal Angle   19
    To Summarize   20
  Cutters   21
    Steel-Tooth   21
    Tungsten Carbide Inserts   22
  Bit Gauge   22
    To Summarize   24
  Drilling Fluid and Hydraulics   25
    Reactive and Nonreactive Materials 26
    Air and Gas Drilling Fluids   26
    To Summarize   27
  Watercourses and Jet Nozzles   28
    Jet Nozzles   28
    Hydraulic Horsepower   30
    To Summarize   30
  Bearings   31
    Roller and Ball Bearing Bits   32
    Journal Bearing Bits   34
    Bearing Lubrication   37
    To Summarize   39
  Wear   40
    Cone Wear   40
    To Summarize   43
    Cutter Wear   44
    To Summarize   48
    Bearing Wear   48
    Other Types of Wear   49
    Normal Wear Example   49
    To Summarize   50

Diamond Bits   51
  The Properties of Diamonds   52
  Natural Diamond Bits   54
    Natural Diamond Bit Manufacture and Design   55

Profile   56
Cutters   58
Hydraulics   60
To Summarize   62
Synthetic Diamond Bits   63
PDC Bits   63
PDC Bit Manufacture and Design   64
Profiles   65
Cutters   66
Hydraulics   68
TSP Bits   69
To Summarize   70
Hybrid Bits   71
To Summarize   72
Diamond Bit Wear   73
To Summarize   76

Special-Purpose Bits   77
Roller Cone Bits   77
Fixed-Head Bits   78
To Summarize   80

Bit Performance   81
Formation Properties   81
How Bits Drill   83
To Summarize   84

Bit Selection   85
Roller Cone Bits   85
Diamond Bits   87
To Summarize   88

Weight on Bit, Rotary Speed, and Penetration Rate   89
Roller Cone Bits   90
Diamond Bits   91
To Summarize   92

Bit Classification   93
Roller Cone Bits   93
Diamond Bits   95
To Summarize   98

Dull Bit Grading   99
To Summarize   102

Costs   103
To Summarize   104

Field Operating Procedures   105
Drilling with a Roller Cone Bit   105
Drilling with a Diamond Bit   106
To Summarize   108

Conclusion   109
Glossary   111
Review Questions   127
Answers to Review Questions   135

Figures

1. A bit drills the hole.   1

2. Circulating drilling fluid lifts cuttings.   2

3. Bit cutters can be steel teeth (a), metal buttons (b), diamonds (c), or special compacts (d).   3

4. Undergauge hole   4

5. Many layers of rock occur in the earth.   5

6. Several wells are required to produce a large reservoir.   7

7. A roller cone, or rock, bit   9

8. Each cone rotates on its own axis.   10

9. The row of cutters on one cone intermesh in the space between the cutters of the cone next to it.   11

10. Cones rotate on bearings.   11

11. A drag bit   12

12. Watercourses   12

13. Steel teeth are milled out of the cone.   13

14. Tungsten carbide buttons are inserted into the cone.   13

15. A bit cone has a machined opening into which the bearings fit.   16

16. On-center (a) and off-center (b) cone alignment   17

17. A chisel penetrates rock to remove it.   18

18. A shovel gouges out dirt.   18

19. Journal angle   19

20. Chisel-shaped, cone-shaped, and hemispherical inserts   22

21. Gauge cutters determine the hole's diameter, or gauge.   22

22. Flat tungsten carbide inserts on the gauge row   23

23. Drilling fluid circulation   25

24. One of the three jet nozzles on this bit   28

25. Extended nozzle   29

26. A center jet nozzle   29

27. Roller and ball bearings   32

28. A plain, or journal, bearing in the nose of the cone   33

29. A journal bearing   34

30. Ball bearings hold the cone on the bearing assembly.   35

31. A retaining ring holds the cone on the bearing assembly.   35

32. Mud cools and lubricates unsealed bearings.   37

33. Grease from a built-in reservoir lubricates sealed bearings.   37

34. Skidding, or dragging, flatten the bit's teeth   40

35. Cone erosion   41

36. Cracked cone   42

37. Off-center wear   42

38. Center coring   43

39. Broken inserts   44

40. Broken teeth   45

41. Chipped cutters   45

42. Flat-crested wear   46

43. Self-sharpening wear   46

44. Tracking wear   47

45. Heat checking   47

46. Gauge rounding   47

47. Bit damaged by junk   49

48. Washout   49

49. Normally worn bit   49

50. A diamond bit has three main parts.   55

51. Single-cone profile   57

52. Double-cone profile   57

53. Parabolic profile   57

54. Concave profile   57

55. Grid plot   58

56. Circle plot   59

57. Ridge plot   59

58. Radial flow watercourses   60

59. Feeder-collector, or cross-pad, watercourses   61

60. A polycrystalline diamond compact (PDC)   63

61. Short parabolic profile   65

62. Shallow-cone profile   65

63. Parabolic profile   65

64. PDC cutters range in size from ⅜ inch to 2 inches (9 to 50 millimetres).   66

65. Two PDC cutter shapes are the cylinder and the stud.   66

66. Back rake angle   67

67. Side rake angle   67

68. PDC cutters in the body of the bit   68

69. Jet nozzles on a PDC bit   68

70. TSPs are triangular or round in shape.   69

71. Diamond-impregnated pad, or stud, on bit gauge surface   71

72. A diamond-impregnated backup stud behind a PDC cutter   72

73. Bottomhole pattern caused by bit whirl and normal rotation   73

74. A spiral-shaped hole   74

75. PDC cutter wear and failures   75

76. A jet deflection bit   77

77. An antiwhirl bit   78

78. An eccentric bit   79

79. A core bit   79

80. A sidetracking bit   79

81. Drilling action of a roller cone bit   83

82. Drilling action of a natural diamond bit   83

83. Drilling action of a PDC bit   83

84. Drilling action of a TSP bit   83

85. Ring gauge held so that two cones touch the ring   101

86. Roller cone vs PDC bits   104

Tables

▼
▼
▼

1. IADC classification for steel tooth and insert bits   94

2. IADC classification for PDC bits   96

3. IADC classification for TSP and natural diamonds   97

4. IADC dull grading chart   100

Foreword

For many years, the Rotary Drilling Series has oriented new personnel and further assisted experienced hands in the rotary drilling industry. As the industry changes, so must the manuals in this series reflect those changes.

The revisions to both text and illustrations are extensive. In addition, the layout has been "modernized" to make the information easy to get; the study questions have been rewritten; and each major section has been summarized to provide a handy comprehension check for the student.

PETEX wishes to thank industry reviewers—and our readers—for invaluable assistance in the revision of the Rotary Drilling Series. Also, we wish to thank the International Association of Drilling Contractors (IADC) for their endorsement of the project. On the PETEX staff, Deborah Caples designed the layout; Doris Dickey proofread innumerable versions; and Ron Baker served as content editor for the entire series. Kathy Bork did her usual superlative job in editing the material.

Although every effort was made to ensure accuracy, this manual is intended to be only a training aid; thus, nothing in it should be construed as approval or disapproval of any specific product or practice.

Sheryl Horton

Acknowledgments

Special thanks to Ken Fischer, director, Committee Operations, International Association of Drilling Contractors, who reviewed this manual and secured other reviewers. John Altermann, Reading & Bates Drilling Company and Jim Arnold, Salem Investment, provided invaluable suggestions on the content and language. Several bit manufacturers also offered tremendous assistance with brochures, technical information, and encouragement. These were Hughes Christensen, Hycalog, Reed, Smith International, and DBS Baroid. Jonell Clardy clarified the sometimes difficult text with wonderful new drawings.

Sheryl Horton

Units of Measurement

▼
▼
▼

Throughout the world, two systems of measurement dominate: the English system and the metric system. Today, the United States is almost the only country that employs the English system.

The English system uses the pound as the unit of weight, the foot as the unit of length, and the gallon as the unit of capacity. In the English system, for example, 1 foot equals 12 inches, 1 yard equals 36 inches, and 1 mile equals 5,280 feet or 1,760 yards.

The metric system uses the gram as the unit of weight, the metre as the unit of length, and the litre as the unit of capacity. In the metric system, for example, 1 metre equals 10 decimetres, 100 centimetres, or 1,000 millimetres. A kilometre equals 1,000 metres. The metric system, unlike the English system, uses a base of 10; thus, it is easy to convert from one unit to another. To convert from one unit to another in the English system, you must memorize or look up the values.

In the late 1970s, the Eleventh General Conference on Weights and Measures described and adopted the Système International (SI) d'Unités. Conference participants based the SI system on the metric system and designed it as an international standard of measurement.

The *Rotary Drilling Series* gives both English and SI units. And because the SI system employs the British spelling of many of the terms, the book follows those spelling rules as well. The unit of length, for example, is *metre*, not *meter*. (Note, however, that the unit of weight is *gram*, not *gramme*.)

To aid U.S. readers in making and understanding the conversion to the SI system, we include the following table.

English-Units-to-SI-Units Conversion Factors

Quantity or Property	English Units	Multiply English Units By	To Obtain These SI Units
Length, depth, or height	inches (in.)	25.4	millimetres (mm)
		2.54	centimetres (cm)
	feet (ft)	0.3048	metres (m)
	yards (yd)	0.9144	metres (m)
	miles (mi)	1609.344	metres (m)
		1.61	kilometres (km)
Hole and pipe diameters, bit size	inches (in.)	25.4	millimetres (mm)
Drilling rate	feet per hour (ft/h)	0.3048	metres per hour (m/h)
Weight on bit	pounds (lb)	0.445	decanewtons (dN)
Nozzle size	32nds of an inch	0.8	millimetres (mm)
Volume	barrels (bbl)	0.159	cubic metres (m³)
		159	litres (L)
	gallons per stroke (gal/stroke)	0.00379	cubic metres per stroke (m³/stroke)
	ounces (oz)	29.57	millilitres (mL)
	cubic inches (in.³)	16.387	cubic centimetres (cm³)
	cubic feet (ft³)	28.3169	litres (L)
		0.0283	cubic metres (m³)
	quarts (qt)	0.9464	litres (L)
	gallons (gal)	3.7854	litres (L)
	gallons (gal)	0.00379	cubic metres (m³)
	pounds per barrel (lb/bbl)	2.895	kilograms per cubic metre (kg/m³)
	barrels per ton (lb/tn)	0.175	cubic metres per tonne (m³/t)
Pump output and flow rate	gallons per minute (gpm)	0.00379	cubic metres per minute (m³/min)
	gallons per hour (gph)	0.00379	cubic metres per hour (m³/h)
	barrels per stroke (bbl/stroke)	0.159	cubic metres per stroke (m³/stroke)
	barrels per minute (bbl/min)	0.159	cubic metres per minute (m³/min)
Pressure	pounds per square inch (psi)	6.895	kilopascals (kPa)
		0.006895	megapascals (MPa)
Temperature	degrees Fahrenheit (°F)	$\dfrac{°F - 32}{1.8}$	degrees Celsius (°C)
Thermal gradient	1°F per 60 feet	—	1°C per 33 metres
Mass (weight)	ounces (oz)	28.35	grams (g)
	pounds (lb)	453.59	grams (g)
		0.4536	kilograms (kg)
	tons (tn)	0.9072	tonnes (t)
	pounds per foot (lb/ft)	1.488	kilograms per metre (kg/m)
Mud weight	pounds per gallon (ppg)	119.82	kilograms per cubic metre (kg/m³)
	pounds per cubic foot (lb/ft³)	16.0	kilograms per cubic metre (kg/m³)
Pressure gradient	pounds per square inch per foot (psi/ft)	22.621	kilopascals per metre (kPa/m)
Funnel viscosity	seconds per quart (s/qt)	1.057	seconds per litre (s/L)
Yield point	pounds per 100 square feet (lb/100 ft²)	0.48	pascals (Pa)
Gel strength	pounds per 100 square feet (lb/100 ft²)	0.48	pascals (Pa)
Filter cake thickness	32nds of an inch	0.8	millimetres (mm)
Power	horsepower (hp)	0.7	kilowatts (kW)
Area	square inches (in.²)	6.45	square centimetres (cm²)
	square feet (ft²)	0.0929	square metres (m²)
	square yards (yd²)	0.8361	square metres (m²)
	square miles (mi²)	2.59	square kilometres (km²)
	acre (ac)	0.40	hectare (ha)
Drilling line wear	ton-miles (tn•mi)	14.317	megajoules (MJ)
		1.459	tonne-kilometres (t•km)
Torque	foot-pounds (ft•lb)	1.3558	newton metres (N•m)

Introduction

▼
▼
▼

A bit is the business end of a rotary drilling rig (fig. 1). As you learned in Lesson 1, *The Rotary Rig and Its Components*, a rig is a large, complex piece of machinery. It has many parts and requires several qualified persons to run it. Keep in mind, however, that all those rig parts have one main purpose: to put a bit on the bottom of a hole and turn it to the right.

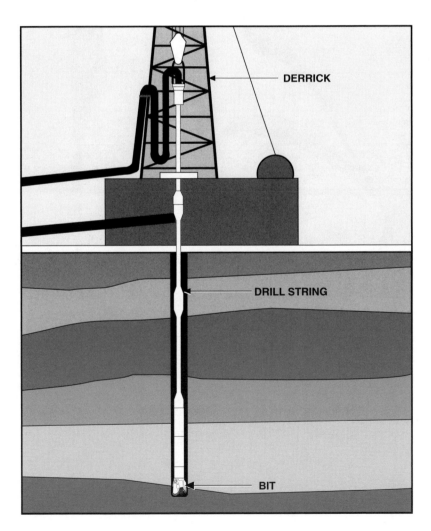

Figure 1. A bit drills the hole.

Besides rotating the bit, the rig crew also has to circulate drilling fluid through it to lift the drilled pieces of rock (the cuttings) out of the hole (fig. 2). Removing the cuttings keeps them from interfering with the bit's cutters. Cutters are the part of the bit that actually cuts, or drills, the rocks. Bit cutters can be steel teeth, very hard metal buttons, diamonds, or specially designed components called compacts (fig. 3).

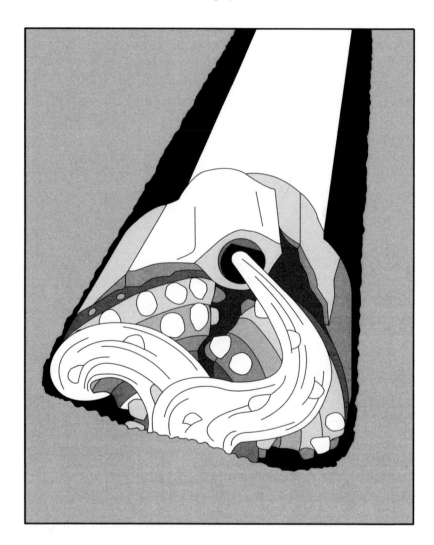

Figure 2. Circulating drilling fluid lifts cuttings.

(a.) steel teeth

(b.) metal buttons

(c.) diamonds

(d.) special compacts

Figure 3.   Bit cutters

ROP, Longevity, and Gauge

▼
▼
▼

Rig owners and operators want a bit that gives a good *rate of penetration*, or ROP (pronounced "are-oh-pee"). They also want the bit to have *longevity*—they want it to last a long time.

What is more, they want a bit that drills a full-gauge hole. *Full-gauge hole* is a hole whose diameter is the same size as the manufactured diameter of the bit. For example, a 7⅞-inch (200-millimetre) bit has drilled a full-gauge hole if the hole's diameter also measures 7⅞ inches (200 millimetres). Rig owners and operators want the bit to drill full-gauge hole during the entire time it is in use. If the sides of the bit wear down, it will drill an undersize, or *undergauge*, hole (fig. 4). Full-gauge bits and other tools rig crew members may lower into an undergauge hole can get stuck. Crew members therefore usually have to open, or *ream*, the under-gauge hole to make it full gauge. Reaming undergauge hole wastes valuable time.

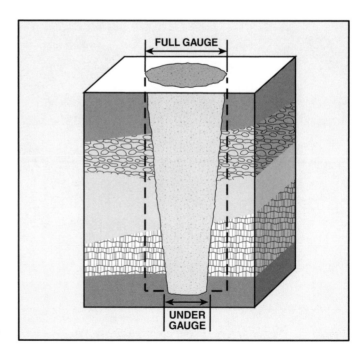

Figure 4. Undergauge hole

4

Formations and Bit Design

The consideration that most affects bit design is the type of rock, or formation, the bit must drill. Is the formation hard? Soft? Abrasive? Sticky? If rock formations were all the same, the driller could select one bit and drill ahead at the maximum rate. In reality, formations consist of alternating layers of soft and hard, abrasive and nonabrasive, rocks (fig. 5), so manufacturers make various types of bits for drilling different formations.

Hardness is a rock's resistance to scratching. Bit cutters must be harder than the rock to scrape or gouge it. Scraping and gouging are intense forms of scratching. An abrasive rock wears and erodes a tool that is working on it. An abrasive rock also wears a tool faster than a nonabrasive rock does, much as a sidewalk dulls a sharpened pencil faster than paper does.

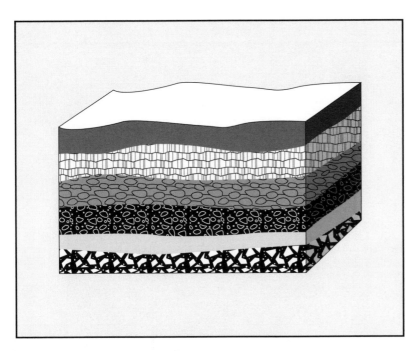

Figure 5. Many layers of rock occur in the earth.

Even though many types of formation exist, it is not practical for rig operators to change bits every time they encounter a different formation. For instance, a particular formation may be made up of mostly medium-soft rocks with shallow streaks of a hard, abrasive material running through it. In this case, the rig operator would probably select a bit designed to drill medium-soft rocks even though the hard streaks may shorten the bit's life. An operator must choose a bit that represents a compromise—that is, one that performs reasonably well under all conditions it might meet.

Choosing a Bit

Choosing the right bit may involve some guesswork when a company drills a *wildcat well*—the first well drilled in a particular area. Nevertheless, the *operator*—the oil company that hires a drilling contractor to drill—does not fly completely blind. The oil company's geologists usually have a good idea of the formations in a particular area. They give this information to the operator or the drilling contractor, who then selects bits that can drill the expected formations.

If the wildcat well strikes a significant amount of oil or gas, the company will probably drill more wells to produce the field. Oil and gas *reservoirs*—formations that contain hydrocarbons—often cover several acres (or hectares) underground and may be several feet (or metres) thick. One well usually cannot drain all the oil and gas from such a large area (fig. 6). Choosing the right bits is easier when drilling additional wells in the field because the operator knows what formations to expect and which bit drills them best.

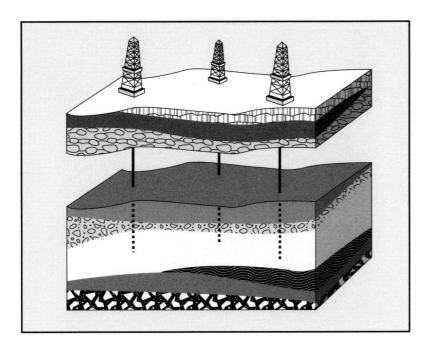

Figure 6. Several wells are required to produce a large reservoir.

Cost also influences which bit to use. Some bits are very expensive—a bit may cost as much as, or even more than, a luxury automobile. Expensive bits usually last longer, however, or drill a particular type of rock better. Crew members do not have to replace such bits as often, so they require fewer time-consuming *trips* (removing and replacing the drill string from the drilled hole). Costs are discussed in more detail on page 103.

To summarize—

- A bit should give a good rate of penetration (ROP), have longevity, and drill a full-gauge hole.
- Undergauge hole is bad because full-gauge bits and other tools can get stuck in it.
- Manufacturers make bits to drill formations of various hardnesses.
- Operators usually choose a bit that represents a compromise—one that performs reasonably well under all conditions it might meet.
- The type of rocks to be drilled and the cost of the bits needed to drill them determine the bits selected to drill a particular well.

Roller Cone Bits

▼
▼
▼

A rock bit is the same thing as a roller cone bit (fig. 7). Oil people called the first roller cone bits "rock bits" because they used them mainly to drill hard rocks instead of the relatively soft dirts and rocks that conventional bits of the time drilled. A roller cone bit is so called because it has three (or sometimes two) hollow cone-shaped components. These metal cones have rows of cutters on the surface.

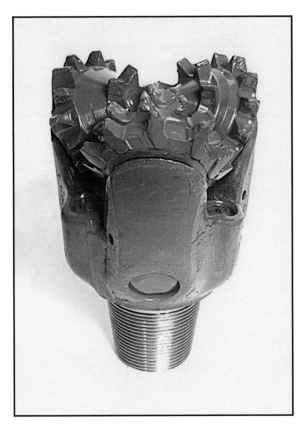

Figure 7.   A roller cone, or rock, bit

Each cone rotates, or rolls, on its own axis (fig. 8), and together all the cones rotate as the drill string rotates. Putting weight on the bit and then rotating it makes the cutters scrape, gouge, or crush the formation to remove it.

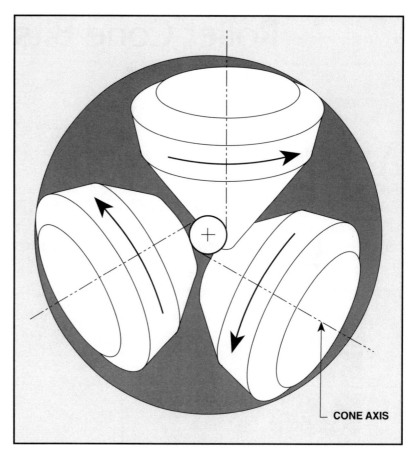

CONE AXIS

Figure 8. Each cone rotates on its own axis.

The cutters of the cones intermesh with each other like gears—that is, one cone's cutters fit into the spaces between another cone's cutters (fig. 9). This intermesh allows the cutters on one cone to clean out any rock or clay that may stick to the cutters on another cone. Keeping the cutters clean helps them drill efficiently. Each cone rotates on bearings that drilling fluid or a special grease sealed inside the cone lubricates (fig. 10).

Figure 9. The row of cutters on one cone intermesh in the space between the cutters of the cone next to it.

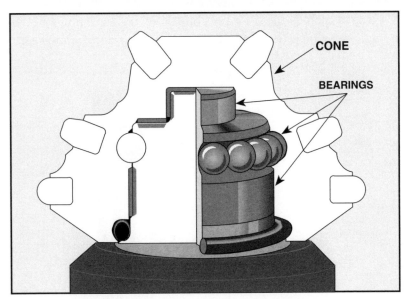

Figure 10. Cones rotate on bearings.

History

The first roller cone, or rock, bit came into use in 1909. Before that, drag bits were the only bits available. A *drag bit* is just two, three, or four steel blades attached to a shank (fig. 11). People called them fishtail bits because the blades looked like the tail of a fish. Fishtail bits worked well in soft formations but not so well in hard rock. As a result, the rock bit steadily gained acceptance for use primarily in hard-formation areas.

Early roller cone bits had only two cones, which did not mesh, so they readily balled up in soft shales. A bit balls up when cuttings mix with the drilling fluid and form a sticky mass on the bottom of the bit. This mass packs into the spaces between the teeth. In effect, balling up makes the cutting edges on the ends of the teeth so short that they hardly penetrate into the formation. A balled-up bit thus has a slow rate of penetration.

In the first bits, the *watercourses*—the passageways drilled through the bit so that drilling fluid can get out (fig. 12)—washed out (eroded) quickly. With the watercourses washed out, the fluid could not clean the bit's cutters. If the erosion was severe, the bit came apart. Abrasive material such as sand in the fluid was the main culprit. Manufacturers solved the erosion problem by fitting a replaceable washpipe inside the watercourses. A *washpipe* is a short pipe with a very hard inside coating that resists erosion.

The three-cone rock bit appeared during the 1930s. It featured bearings lubricated by the drilling fluid and different types of teeth for different formations. These bits appear outwardly to be quite similar to those available today. Today's bits, however, have much better design, metallurgy (the study and use of metals), and manufacture.

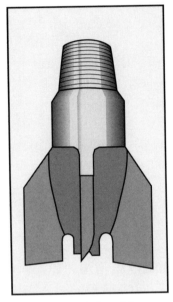

Figure 11. A drag bit

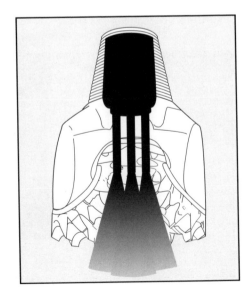

Figure 12. Watercourses

Today, two general types of roller cone bits are available: *steel-tooth bits* (also called *milled-tooth bits*) and *tungsten carbide insert bits*. Steel-tooth bits have teeth that the manufacturer mills, or cuts, out of the body of the cone after the cone is cast (fig. 13). Some manufacturers forge the teeth from the cone. Forging uses a heavy-duty press to compress the cone metal and form the teeth: no cutting, or milling, is involved.

Types of Roller Cone Bits

Figure 13. Steel teeth are milled out of the cone.

Tungsten carbide insert bits (often called insert, or button, bits) are named for the tungsten carbide inserts in the cones. The inserts, or buttons, are small solid cylinders that have rounded or softly sharpened ends. The manufacturer presses, or inserts, the buttons into holes drilled in the cone (fig. 14). Both teeth and inserts are called cutters.

Tungsten Carbide Bits

Figure 14. Tungsten carbide buttons are inserted into the cone.

13

Tungsten carbide is a gray metal powder that manufacturers heat with a special binder and cast in a mold to make inserts. They blend the binder with the tungsten carbide powder and place the blend in the mold; they then heat the mixture to melt the binder, which holds (binds) the powder in place. Manufacturers can also apply tungsten carbide in layers to steel teeth and other parts of a bit that may contact the side of the hole. Since tungsten carbide is much harder than steel, it withstands abrasion better and increases the life of a bit. Applying tungsten carbide to a steel-tooth bit does not, however, make it a tungsten carbide insert bit. Only bits with tungsten carbide inserts, instead of steel teeth, are insert bits.

Insert bits have some advantages over milled-tooth bits. Tungsten carbide wears very little, so the inserts last longer than steel teeth. Also, the same tungsten carbide bit can drill many different types of formations, so the driller does not need to change bits as often.

Disadvantages include the possibility that, as the inserts dig deeply into the rock, the cones in which the buttons are mounted can hit the formation and transmit impact shocks to the bearings. These shocks can destroy the bearings and ruin a bit. Another disadvantage is that abrasive drilling fluid can erode the cones so severely that the inserts fall out. What is more, tungsten carbide insert bits are much more expensive than steel-tooth bits and cannot drill as fast as a steel-tooth bit in most soft-to-medium-hard formations.

Roller Cone Bit Manufacture and Design

The best designs for machinery often result from a series of compromises. Take the automobile, for example. A large car has a long wheel base for smooth riding, wide tire tread for stability, a wide body for roominess, and abundant power for speed and acceleration. On the other hand, a smaller car offers other desirable characteristics—easy parking, fuel economy, and a low-compression engine that uses inexpensive low octane gas.

But one car cannot include all of these desirable features. To get better fuel economy, the car must be lighter in weight. To be lighter, however, it might have a small engine and therefore be less powerful. Or it might be made of a lighter, thinner metal and therefore not be as safe in an accident. The designer must decide where to compromise so as not to eliminate completely one desirable feature while emphasizing another.

Similarly, a bit designer wants heavy-duty bearings for long life, a strong cone that will not wear away, and cutters sized for a fast rate of penetration. Each feature competes with the others for the limited space on a bit.

In addition, a roller cone bit designer must choose whether to make the cutters out of steel or tungsten carbide. Tungsten carbide is harder than steel but more brittle. Steel is softer than tungsten carbide but more pliable. A brittle material does not break easily when compressed, even by a great force, but it does break under an impact, or a sudden blow. In drilling, impacts occur, for instance, when the bit suddenly hits a ledge of hard rock. This can cause the bit to bounce, producing impact shocks to the cutters.

Cones Each cone is a hollow part cast in a mold (fig. 15). It has teeth or inserts on the outside, and the bearing assembly fits inside the hollow and attaches the cone to a leg of the bit. How thick to make the cone is one of the compromises the designer must make. A thicker cone is stronger, but it leaves less room for the bearings. The bearings wear an enormous amount because of the weight on the bit and the speed of rotation. The designer therefore must make the cone thin enough to allow enough space for heavy-duty bearings.

The cones are made of a steel alloy. An *alloy* is a mixture of metals. All steels are iron and carbon with small amounts of other metals added to improve heat resistance and hardness. Some of the preferred alloys contain high percentages of nickel and molybdenum. Harder alloys can also include chromium, silicon, or cobalt.

Bit manufacturers make steel stronger by treating the finished parts of the bit with heat. This heating process consists of three steps:

1. Carburizing—heating the steel in a carbon-rich atmosphere. In this way the outer layer absorbs carbon, which forms a tough outer shell.

2. Heating and quenching—immersing the hot carburized steel in a liquid so that the steel cools quickly. This action causes the carbon atoms to become trapped between the iron atoms. Heating and quenching makes the steel harder but more brittle.

3. Tempering—reheating to a lower temperature than in step 2. Tempering prevents cracks from forming due to the stresses that quenching causes. It also restores the steel's impact strength.

Figure 15. A bit cone has a machined opening into which the bearings fit.

When assembling the cones on a bit, the manufacturer can either align each cone so that they do not line up with the center axis of the bit, or align each cone so that they do line up with the center axis of the bit. Cone alignment is thus either *on-center* (fig. 16a) or *off-center* (fig. 16b).

Cone Alignment

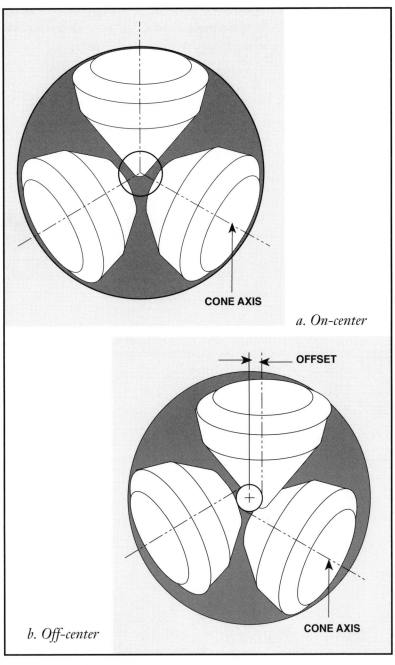

a. On-center

b. Off-center

Figure 16. Cone alignment

If the cones have on-center alignment, the cutters rotate around a common center as the bit rotates. Rotating around a common center causes the tip of the cutters to contact the formation straight-on. This straight-on contact, along with weight on the bit, forces the cutters straight into the rock. The cutters punch into the rock and fracture it, much like a pointed chisel penetrates and removes rock when struck with a hammer (fig. 17). Just as the hammer's impact forces the chisel to penetrate the rock, the weight on the bit forces the cutters to penetrate the formation.

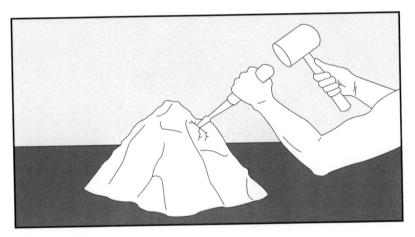

Figure 17. A chisel penetrates rock to remove it.

Off-center alignment, called offset, changes the way the cutter tips contact a formation. Offset cutters do not rotate around a common center as the bit rotates. Offset causes the cutter tips to drag, or scrape, across the formation. Offset makes the teeth act much like a shovel: the teeth scrape across the formation and gouge it out (fig. 18).

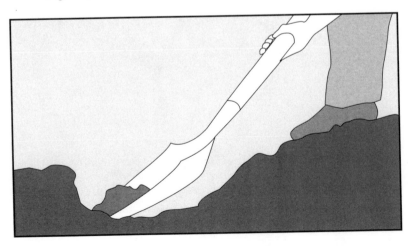

Figure 18. A shovel gouges out dirt.

The scraping and gouging action caused by offset is best for soft formations because scraping across soft rocks is the best way to remove soft material. Think about hammering on a pointed chisel in soft dirt. The chisel deeply penetrates the dirt, but it does not remove any.

Bits for medium formations have some cone offset but not as much as those for soft formations. Medium-formation bits combine scraping, gouging, and penetration of the formation.

Interfit, or *intermesh*, is the distance that the ends of the cutters of one cone extend into the spaces of the adjacent cone. As described before, interfit mechanically cleans the spaces between the rows of cutters. Good interfit also allows the cones to be larger, since the teeth or inserts do not meet tip to tip but extend into spaces in the adjacent cone. Larger cones, in turn, allow longer cutters, thicker cones, and more space for the bearings.

Interfit

The *journal angle*, or the *pin angle*, is the angle formed by the center line of the bearing pin and the horizontal plane (fig. 19). This angle plays a role in the size of the cones. Each cone must completely fill the allotted space to intermesh correctly and drill the proper size hole. So the larger the journal angle, the larger the diameter of the cone (and therefore the bit) can be.

Journal Angle

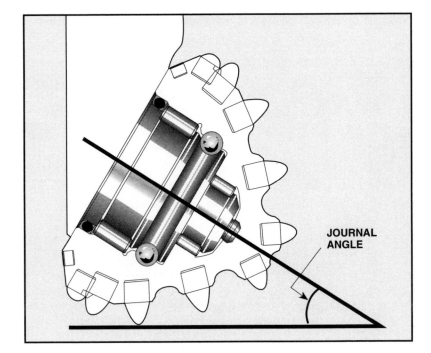

JOURNAL ANGLE

Figure 19. Journal angle

To summarize—

Two types of roller cone bits—
- Steel-tooth bits
- Tungsten carbide insert bits

Characteristics of roller cones—
- Cast, hollow, metal pieces inside of which bearings fit and, if they are on steel-tooth bits, out of which steel teeth are milled or forged. In tungsten carbide bits, tungsten carbide inserts are inserted into the outer surface of each cone.
- May be aligned on-center or off-center.
- Interfit mechanically cleans the spaces between the rows of cutters.
- Journal, or pin, angle plays a role in the size of the cones.

Steel-tooth bits have long or short teeth, depending on the hardness of the formation they will drill. Long teeth are for soft formations, short teeth are for hard formations, and medium-long or medium-short teeth are for medium-soft or medium-hard formations, respectively.

In soft formations, where the tooth scrapes and gouges the formation, long teeth are desirable because they can remove a lot of formation. Also, long teeth tend not to break in soft formations. In hard formations, on the other hand, where the tooth punches into the formation, short teeth are desirable because they shatter the rock without breaking themselves. A long tooth cannot absorb as much impact as a short tooth without breaking. (It is like driving a pole into hard ground. If a lot of the pole sticks out of the hole you drove it into, it is easy to break the pole. If, on the other hand, not very much of the pole sticks out of the hole, then it is harder to break.) For medium formations, tooth lengths fall somewhere between the longest and the shortest.

Manufacturers cover the teeth of some steel-tooth bits with tungsten carbide to minimize wear in hard, abrasive formations. Such steel-tooth bits are not insert bits, however. They still have milled teeth instead of inserts as the main cutting structure.

Some steel-tooth bits have self-sharpening teeth. The manufacturer puts tungsten carbide hardfacing on only one side of the tooth. The side of the tooth without the hardfacing wears faster than the hardened side, thus enabling the hardened side to hold a sharp edge.

Cutters
Steel-Tooth

Tungsten Carbide Inserts   Insert bits have chisel-shaped, cone-shaped, or hemispherical cylinders made of tungsten carbide (fig. 20). The shape of the inserts determines how hard a formation the bit can drill. The longer chisel-shaped inserts are good for softer rock, where they scrape and gouge. Cone-shaped inserts are for medium formations, where they combine scraping and gouging with penetration. Short hemispherical inserts are good for the hardest rock, where they penetrate and shatter the formation.

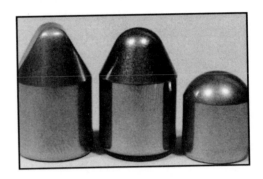

Figure 20. Chisel-shaped, cone-shaped, and hemispherical inserts

Bit Gauge   Gauge can refer to the diameter of the hole or to the diameter of the bit used to drill it. The gauge of the bit is also any part of the bit that contacts the sides of the hole.

The parts of the bit, on both the cones and the flank, that parallel the wall of the drilled hole are called the *gauge areas*. The *gauge cutters* (also called the *gauge row* or the *heel row*) are the outermost row of cutters. These gauge cutters cut the outside edge of the hole and therefore determine its diameter (fig. 21). Changing the angle of the gauge cutters on the surface of the cone increases or decreases the diameter of the bit without changing the size of the cones.

Figure 21. Gauge cutters determine the hole's diameter, or gauge.

Unlike the rest of the rows, the gauge rows of each cone do not intermesh. To reduce a bit's balling up, this row may have missing teeth in the pattern to create more space between the cutters.

The gauge cutters are very important because they determine the size of the hole. They also take the most stress because they are the part of the bit that first drills the rock. The gauge area also wears because of abrasion from the drilling fluid and cuttings. To reinforce these areas, steel-tooth bits have a tungsten carbide coating on the gauge of the cone or tungsten carbide inserts alongside the gauge teeth (fig. 22). These inserts relieve the cutters of part of the impact load and wear in very hard or abrasive formations. This helps the gauge cutters to maintain the gauge of the hole.

Figure 22. Flat tungsten carbide inserts on the gauge row

To summarize—

Characteristics of steel-tooth bits—
- Long teeth for soft formations
- Short teeth for hard formations
- Medium-long or medium-short teeth for medium-soft or medium-hard formations

Shapes of tungsten carbide inserts—
- Hemispherical for hard formations
- Cone-shaped for medium formations
- Chisel-shaped for soft formations

Bit gauge areas include—
- Sides of cones that contact the wall of the hole
- Sides of bit that parallel the wall of the hole

Gauge cutters—
- the outermost row of cutters

In drilling, *hydraulics* refers to the science of drilling fluid circulation. The drilling fluid circulates from aboveground, through the drill string, out of openings in the bit, and back up the hole through the annulus (fig. 23). Drilling fluid is usually a liquid, but operators also use compressed air, compressed natural gas, or foam. A drilling fluid that is liquid is also called *drilling mud*. Drilling mud has three main jobs:

1. It carries the cuttings up the hole and away from the bit. This action keeps the bottom of the hole clean so that the bit is touching only virgin formation and is not just redrilling the cuttings.

2. It cools the bit, which heats up because of friction.

3. Its weight develops pressure. Drilling mud pressure offsets the pressure of the formation. If the pressure of the drilling mud is lower than the pressure in the formation, then any fluids in the formation, such as gas, oil, or salt water, could enter the borehole. Entry of formation fluids into a hole is a kick. If the crew fails to detect the kick, or if it does not take the proper steps to control the kick, then the well could blow out. A blowout is the uncontrolled flow of formation fluids from the hole.

Drilling Fluid and Hydraulics

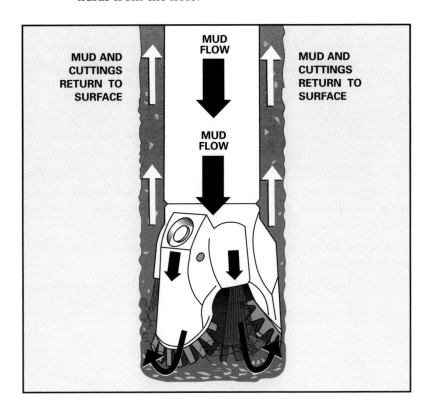

Figure 23. Drilling fluid circulation

25

*Reactive and
Nonreactive Materials*

Drilling mud is usually a mixture of water or oil, clays, a weighting material to make it dense or heavy, and a few chemicals. The chemicals and solid materials added to the water or oil are either *reactive* or *nonreactive*. Reactive materials react chemically with the mud's liquid or other components. Nonreactive materials do not react with the mud's other components.

A reactive material, such as bentonite, when added to fresh water, makes the mud gel when it stops moving. This gelling holds the cuttings suspended wherever they are, like fruit in gelatin. A mud without bentonite or another reactive material does not gel and allows the cuttings to fall to the bottom of the hole when circulation stops.

Nonreactive materials also give the mud desirable properties, even though they do not react with the liquid or other chemicals in the mud. One widely used nonreactive material is barite (barium sulfate—$BaSO_4$). Because it weighs over four times more than water, crew members add barite to mud to make the mud heavy or dense. A heavy mud offsets formation pressure to prevent kicks.

*Air and Gas Drilling
Fluids*

Air and natural gas circulate in the same way as liquids. Their advantage over liquids is that they blow the cuttings out of the hole faster because the lighter weight of the gas does not hold the cuttings down. This action speeds up the rate of penetration and makes drilling more efficient.

The main drawback with air drilling has to do with the water that occurs naturally in many formations. If enough water is present, it flows into the borehole during drilling, mixes with the fine cuttings produced by air drilling, and creates large, sticky masses that ball up the bit and block circulation. As a result, drilling stops.

If, however, not too much water is present, well drillers can add a surface active agent, a *surfactant*, to the air or gas, which causes the water to foam. The foam keeps the cuttings from balling up and carries them out of the hole. In any case, drilling with air, gas, or foam can give remarkable penetration rates.

Both milled and insert bit designs for circulating air, gas, foam, and liquid are available.

To summarize—

Jobs of drilling fluid —

- It removes cuttings from the bottom of the hole and carries them to the surface for disposal
- It cools the bit
- It develops pressure to offset formation pressure

Basic composition of drilling mud—

- Water or oil
- Reactive materials
- Nonreactive materials
- Chemicals

Characteristics of air, gas, or foam drilling—

- Circulate the same way as liquids but remove cuttings better because they are lighter and do not hold the cuttings down.
- Large quantities of formation water can wet the cuttings and form large globs of sticky matter that clog up the hole and prevent circulation.

Watercourses and Jet Nozzles

The drilling fluid comes down the drill string and out of the bit through either watercourses or jet nozzles. *Watercourses*, the older design, are just holes drilled in the center of the bit that allow the fluid to flow out. They direct the fluid onto the cutters to clean and cool them.

In many cases, however, drilling fluid coming out of watercourses does not adequately clean the cutters or the bottom of the hole. *Hole cleaning*—moving cuttings away from the bit's cutters—is important to the rate of penetration. If the drilling fluid exiting the bit does not adequately clean the hole, then the bit's cutters redrill old cuttings and the rate of penetration suffers.

The speed, or velocity, of the mud as it leaves the bit influences how well the drilling mud cleans the cutters and the bottom of the hole. In bits with watercourses, the velocity of the mud leaving the bit is relatively slow—say 100 feet (30 metres) per second, or about 68 miles (about 110 kilometres) per hour. In some cases, this speed is adequate to clean the hole; in many cases, however, it is not.

Jet Nozzles

Because watercourses often do not effectively clean the bottom of the hole, bit designers came up with jet nozzles. Just as water coming out of a garden hose increases in speed if you use a nozzle on the end of the hose, jet nozzles on a bit create high-velocity jets of drilling fluid as the fluid leaves the nozzles. This velocity increase creates turbulence around the bit, which cleans and cools the cutters and sweeps the cuttings away. Most modern roller cone bits have three jet nozzles on the side of the bit between the legs (fig. 24). The nozzles are made of a special erosion-resistant material to minimize wear from the fast-moving stream of abrasive drilling mud.

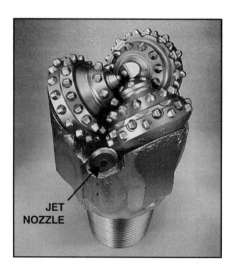

Figure 24. One of the three jet nozzles on this bit

Jet nozzles come in various sizes and are interchangeable. The operator can thus change the size of the nozzles to match the volume of fluid being pumped and keep the jet's speed high. The International Association of Drilling Contractors (IADC) measures nozzles in thirty-seconds of an inch or in millimetres. In the English system, the IADC recommends using only the first number in the fraction to indicate nozzle size. For example, a 10 nozzle is ¹⁰⁄₃₂ inch (7.938 millimetres) in diameter, a 9 nozzle is ⁹⁄₃₂ inch (7.144 millimetres), and so on. With the right combination of fluid volume and nozzle size, fluid velocities in jet bits may exceed 400 feet (122 metres) per second, or over 270 miles (about 440 kilometres) per hour.

On larger bits, the standard position of the jets puts them several inches (centimetres) from the bottom of the hole, so the hydraulic power weakens before the fluid reaches the cuttings. To overcome this problem, manufacturers make bits that have jets that extend from the body of the bit. These elongated jet nozzles, called *extended nozzles* (fig. 25), deliver more hydraulic power to the bottom. Extended nozzles also help prevent balling in soft formations by cleaning the cuttings away better.

Some bits also have a *center jet* along with the three standard side jets (fig. 26). Each cone then has fluid flowing past it on the inside, as well as the outside, which, under certain conditions, may clean the cutters better and reduce balling.

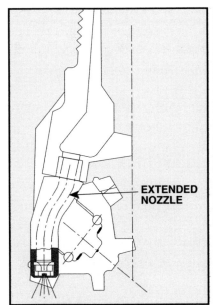
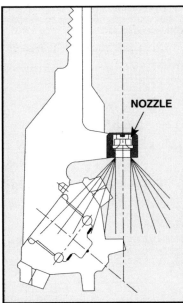

Figure 25. *Extended nozzle (only one of three is shown)*

Figure 26. *A center jet nozzle*

Hydraulic Horsepower

One way to look at jet cleaning power is to express it as hydraulic horsepower. By increasing or decreasing hydraulic horsepower at the bit's jets, the rig operator can vary the hydraulic cleaning power of the jets. A direct relationship exists between hydraulic horsepower and the rate of penetration. The higher the hydraulic horsepower (fluid volume plus velocity), the faster the drilling is, if the driller keeps the correct weight on the bit, rotates the bit at the correct rpm, and if the characteristics of the formation remain the same.

To increase or decrease the amount of hydraulic horsepower delivered by the bit nozzles, the rig crew changes their size. Generally speaking, the smaller the nozzle sizes, the greater is the hydraulic horsepower. The amount of hydraulic horsepower the bit nozzles can deliver is, however, limited by several items. For instance, the high pressures created by small nozzles can increase pump maintenance costs, and the drill pipe, swivel packing, and other rig components may be eroded, or washed out, by the high-pressure drilling fluid. Other limits include (1) the amount of power available to drive the pumps; (2) how much pressure the pumps are designed to deliver normally; (3) the lowest speed needed to return the fluid and cuttings to the surface; and (4) the smallest practical nozzle size (small-diameter nozzles plug more easily than larger sizes).

To summarize—

Jet nozzles—
- Improve hole cleaning
- Are sometimes extended on large bits to improve hole cleaning
- One can also be placed in the center of a bit (in addition to the three standard side jets) to improve hole cleaning

Bit hydraulic horsepower—
- Cleans the bottom of the hole
- Is adjusted at the bit by changing the size of the nozzles

The amount of hydraulic horsepower that the bit nozzles can deliver is limited by—
- High pump maintenance costs
- Drilling fluid erosion of drill pipe and other components
- Pump size and power
- The need for enough annular velocity to lift cuttings
- A small nozzle's tendency to plug easily

Friction is the resistance to the motion of two surfaces rubbing against each other. Friction in tools can be a problem because it slows down movement, heats up surfaces, and wears them away.

Briskly rub your hands together and they get hot. Friction causes the heat. If you press them together hard (add force), it becomes more difficult to rub them, because pressing your hands together hard increases the friction force.

A *bearing* is the device that sits between the cone and its attachment to the leg of the bit to reduce the force of friction as the cone rotates. Roller cone bit bearings must operate under severe conditions. They must withstand the high temperatures that friction produces without *spalling*. Spalling occurs when metal from the bearings flakes off.

Bearings must also continue to move freely in spite of the great stress caused by the weight on the bit. For example, the driller may rotate a bit at 250 rpm with 40,000 pounds (17,800 decanewtons) of weight continuously for 30 to 300 hours.

Bearings provide perhaps the greatest challenge to the designer, who must compromise between the strength needed and the amount of space available for them. Thus, bearings take many forms. A thin band of low-friction material around a moving part is a bearing, but so is a complex assembly of rolling parts and retainers.

Roller cone bits use three types of bearing in the bearing assembly: ball bearings; roller bearings; and journal bearings. Some roller cone bits use all three and some use only roller and ball bearings.

Bearings

*Roller and Ball
Bearing Bits*

A *ball bearing* consists of spheres of metal (balls) inside a track, or groove, called a race. Their rolling action, which replaces the rubbing action, is what reduces friction in both ball and roller bearings. The ball bearings sit next to the roller bearings, closer to the nose of the cone. The requirements for the ball bearing are the same as for the roller bearing—that is, the designer must balance the size and number of balls. Since the ball bearings rotate in a race that the manufacturer machines relatively deeply into the cone and into the leg of the bit, the ball bearings lock the cone on the bit's leg.

A *roller bearing* consists of solid cylinders of metal (rollers) packed side by side into a race. The rollers rotate freely in the race. In this type of bearing assembly, the roller bearings sit inside the widest part of the hollow cone (fig. 27). In figure 27, notice that the bit also uses ball bearings, as well as another set of smaller roller bearings near the end, or *nose*, of the cone. Some roller bearing bits do not use roller bearings in the nose. Instead, they use a plain, or journal, bearing (fig. 28).

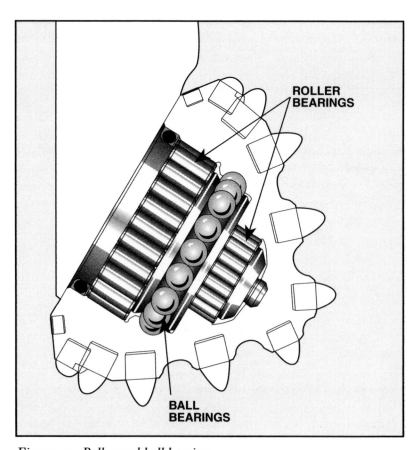

Figure 27. Roller and ball bearings

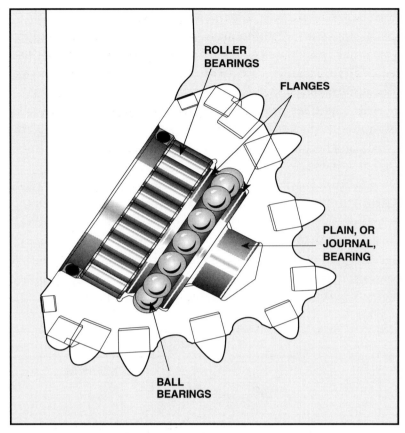

ROLLER
BEARINGS

FLANGES

PLAIN, OR
JOURNAL,
BEARING

BALL
BEARINGS

Figure 28. A plain, or journal, bearing in the nose of the cone

The designer must compromise between the number and the size of rollers. Rollers with the largest diameter will be stronger. The larger their diameter is, however, the smaller the number that will fit into the available space inside the cone. Using fewer rollers increases the load on each one and causes damage such as breakage. Another consideration is the strength and size of the race. The *flanges*, the two rims of the race that hold the rollers in, need to be thick enough so that they will not break.

Journal bearing bits have a journal bearing as the main bearing. Some journal bearing bits also use ball bearings to hold the cone onto the bit; others eliminate the ball bearings and use a retaining ring to hold the cone onto the bit. In all roller cone bits, the inside shape of the cone matches the contour of the bearing assembly. Each bearing in the assembly fits snugly against the cone and takes part of the load.

Journal Bearing Bits

Instead of rollers or balls, a *journal bearing* consists of a flat, polished area around the circumference of the shaft—the *bearing pin*—to which each cone is attached (fig. 29). The highly polished surface reduces friction, and the manufacturer inserts special metal alloys, such as silver and copper, into the part of the cone that contacts the journal. Under the heat and pressure of drilling, these alloys help lubricate the bearing. The heat and pressure cause the metal alloy to melt and become a heavy-duty lubricant between the cone and journal surfaces.

A journal bearing is also called a friction bearing or a plain bearing. In a journal bearing bit, a journal bearing replaces the roller bearings, and, in some, the ball bearings. The largest part of the journal bearing is a cylinder of metal that contacts the inside of the cone along a wide area. As mentioned earlier, in some journal bearing bits, ball bearings hold the cone onto the bearing assembly (fig. 30). In others, a special retaining ring holds the cone (fig. 31).

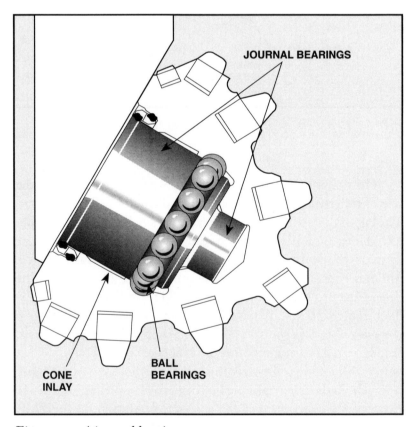

Figure 29. A journal bearing

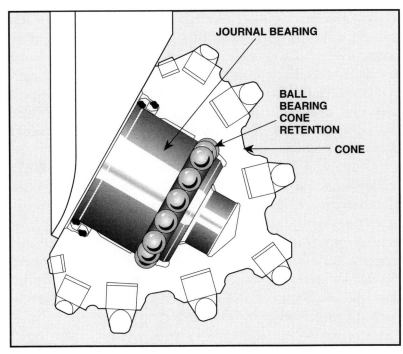

Figure 30. Ball bearings hold the cone on the bearing assembly.

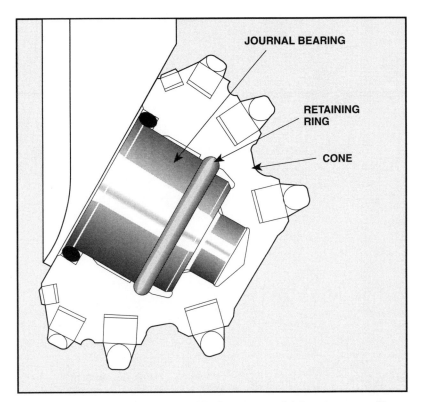

Figure 31. A retaining ring holds the cone on the bearing assembly.

A journal bearing is stronger than a roller bearing because it is a single, large piece of metal that provides surface-to-surface contact over a relatively wide area. Each roller, on the other hand, actually touches the cone only along a thin line parallel to its length as it turns. Think of laying a drinking glass down on its side on a table. Only a small line along its length is touching the table. This line moves as you roll the glass, but it is always a very small part of the whole area of the glass.

Contrast this line contact with a journal bearing, which is like putting your hand flat on a table. The palm of your hand represents a large area touching the friction surface and supporting the load. This large area of contact spreads out the load the bearing must support over a large area. As a result, a journal bearing can last longer than a roller bearing. In fact, journal bearing bits now have virtually unlimited life under normal use if the lubrication system does not fail. Lubrication of the bearing assembly is the key to preventing wear caused by friction, or *galling*.

Lubrication helps rotating metal parts slide smoothly and freely against each other with less friction. A lubricant works like waxing snow skis to make them glide faster. For many years, manufacturers produced rock bits only with nonsealed bearings. Today, most bits have sealed bearings. In a nonsealed bearing, no seal exists between the shirttail of the bit and the outer edge of the cone. Mud enters the bearings and cools and lubricates them (fig. 32).

Unfortunately, drilling mud is not the best lubricant. It often contains abrasive particles drilled from the formation, even though special equipment conditions (cleans) the mud continuously. An unsealed bearing thus has a shorter life than a sealed bearing because abrasive drilling fluid can cause abrasion and galling of all the bearing elements, especially the roller races.

Sealed bearings are, as the name implies, sealed off with a metal or an elastomer (synthetic rubber) ring so that the drilling mud cannot reach them. They therefore need another means of lubrication. The manufacturer puts a reservoir of lubricant in each leg of the bit to feed grease to the bearing assembly in each cone (fig. 33).

Bearing Lubrication

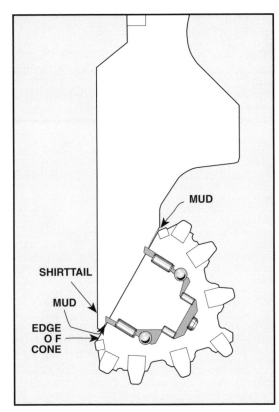

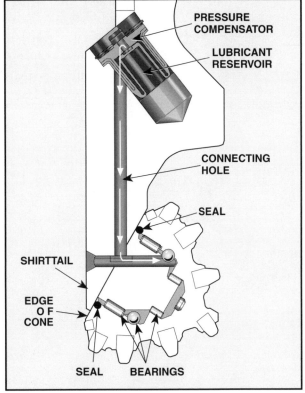

Figure 32. Mud cools and lubricates unsealed bearings.

Figure 33. Grease from a built-in reservoir lubricates sealed bearings.

The reservoir and lubrication system has two functions: (1) it lubricates the bearing and (2) it equalizes the pressure inside the sealed bearing with the pressure in the hole. Equal pressures ensure even distribution of the lubricant over the bearings.

The essential components of a sealed bearing include the seal, a reservoir, and a pressure compensator. The seal keeps the lubricant in and the drilling mud out, while the reservoir holds the lubricant. The manufacturer puts the lubricant in the reservoir, where it should last for the life of the bit. The pressure compensator is a diaphragm that bends as the pressure of the drilling fluid pushes on it. This push equalizes the pressure in the bearing and the pressure of the drilling fluid. Equal pressure allows the lubricant to flow through a passageway to the bearing.

A lubricant may be liquid or solid. The main lubricant used in sealed bearings is a thick grease fed from the reservoir. In addition to grease, some bits also have self-lubricating bearings. They take advantage of friction heat for lubrication. In this type, the inside of the cone has copper inlays imbedded in it. As the bit heats up from friction, the copper melts in small quantities and smears the surface to lubricate it. Meanwhile, grease from the bit's reservoir also lubricates the bearings.

Sealed roller bearings, introduced first in tungsten carbide insert bits and later in steel-tooth bits, have increased bearing life by as much as one-fourth. All major manufacturers of roller cone bits now offer sealed roller and journal bearings as well as nonsealed bearings.

To summarize—

Bearings used in roller cone bits—
- Roller bearings
- Ball bearings
- Journal bearings

Roller bearings are—
- Hard steel cylinders

Ball bearings are—
- Hard steel balls

A journal bearing is a—
- Flat, polished area on the circumference of the bearing pin

A sealed bearing bit has—
- Seals between the cone and the shirttail to keep drilling fluid away from the bearings

On each bit leg, sealed bearing bits have a—
- Seal
- Grease reservoir
- Pressure compensator

Wear

Bits wear out eventually. Ending a run because of unnecessary wear on a bit is, however, costly, because of both the expense of replacing the bit and the time lost. The bit designer is responsible for engineering the best possible combination of materials and design; the driller's responsibility is to use the bit under the appropriate conditions to get the longest life from it. Evaluating how and where the used bit is worn can reveal problems and suggest ways to eliminate or reduce them. It is like looking at the wear of your tire tread to see if the front end of your car is aligned. Bit evaluation can also be a valuable source of information about the formation or drilling conditions. IADC has devised a list of codes for recording the condition of used bits. Called dull bit grading, it is covered on pages 99–102.

The parts that can fail in a roller cone bit are the cones, the cutters, and the bearings. If the penetration rate is slower than expected (given the type of formation and choice of bit, hydraulics, weight, and rotary speed), the operator may have rig crew members pull the bit and examine it for wear. The pattern and type of wear on the cutters, cones, and bearings indicate why the bit is not drilling as it should. Improper weight on the bit, drilling fluid pumped at too high a volume, using the wrong bit, and junk in the hole cause the most damage to bits.

Cone Wear

Cone problems usually result from improper drilling practices or poor cone design. They can be very serious.

Cone skidding or *dragging* occurs when a cone locks (stops turning) as the bit rotates. This flattens the part of the locked cone that contacts the bottom of the hole (fig. 34). Balling up of the bit or *junk* (loose pieces of metal in the hole) lodged between the cones can lock the cones, as can bearing failure or a pinched bit. A pinched bit is a bit in which the bit's legs and cones are forced toward the center of the bit. A driller can pinch a bit by jamming it into an undersized (undergauge) hole. Finally, inadequate break-in can cause skidding (see pages 105–107 for procedures for breaking in a new bit).

Figure 34. Skidding, or dragging, flatten the bit's teeth.

Cone interference is a condition in which the cones are bent inward to the point that they interfere with each other's ability to rotate; in other words, the cones run into each other. Interference can happen when one or more bearings fail and cause a cone to lock and bend inward. Interference can also occur when the driller uses a bit to ream an undergauge hole and puts excessive weight on it. The weight pinches in the cones. Cone interference is not a form of wear, but it is the result of wear to the bearings or bit and causes failures such as cracked cones and broken cutters.

Cone erosion simply means that the cone has worn away. Erosion most often occurs in air or gas drilling, where the drilling fluid velocity is very high and combines with abrasive cuttings to erode the cone, like sandblasting paint off a wall (fig. 35). Erosion can cause the inserts to fall out.

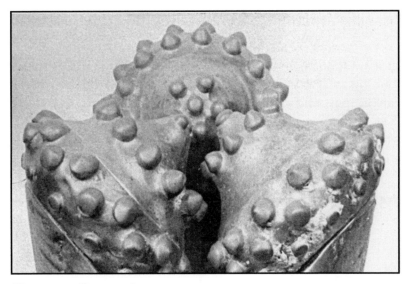

Figure 35. Cone erosion

Cracked cones can occur because of severe erosion (fig. 36). Cones can, however, crack in many other ways. For example, cones may crack where junk in the hole has dented them, where the bit has hit a ledge, or where the cones hit the bottom of the hole hard because the driller dropped the drill string. Also, cone interference, overheating, or chemical corrosion from hydrogen sulfide (H_2S) in the drilling fluid can crack a cone. Defective manufacture or design can also cause cones to crack. Cracks can migrate from the area of the inserts because of the stresses produced by pressing the inserts into the cone during manufacture.

If a cone breaks but most of it remains attached to the bit, you have a *broken cone*. A cone can break because of cone interference or corrosion. Dropping the drill string or hitting the bit on a ledge during a trip can also break a cone or knock it off the bit altogether. When the cone is knocked off the bit, it is said to be lost. It is essential to remove the lost or broken cone pieces from the hole with some type of junk retrieving or milling tool before resuming drilling.

Off-center wear often occurs when the rate of penetration is too slow in a soft or medium-soft formation. Penetration rates that are too slow can occur when the driller fails to rotate the bit at the correct rpm or fails to put enough weight on the bit. When the bit does not drill the hole fast enough, cone offset causes the bit to whirl. *Whirling* is the motion a bit makes when it does not rotate around its center. Instead, it drills with a spiral motion. Whirling causes the bit to drill an overgauge hole (a hole that is too big; the opposite of undergauge). Two cones drill the bottom of the hole and one drills the side. Ridges form on the bottom of the hole and rub against the cones. Off-center wear shows up as grooves between the rows of cutters or extra wear on the gauge area of one cone (fig. 37).

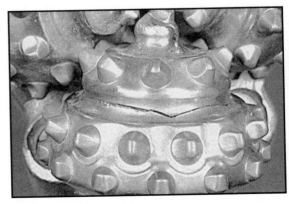

Figure 36. Cracked cone

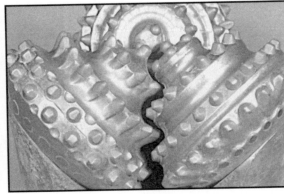

Figure 37. Off-center wear

One remedy for off-center wear is to use a stiff drilling assembly to stabilize the bit. A *stiff drilling assembly* is several large-diameter drill collars and stabilizers run above the bit. A stiff drilling assembly does not bend, or flex, very much as the string rotates. Another remedy is to increase the rate of penetration by using more weight on the bit. Also, it is important to select the proper bit. Keep in mind, however, that sometimes drilling or equipment requirements make it impossible to avoid off-center wear.

Center coring is a condition in which the inside row of cutters on the cones wears or breaks, or the nose of one or more cones breaks off (fig. 38). Junk in the hole, cone erosion, or improper breaking in of the bit can cause center coring. In some cases, a formation too abrasive for the bit causes center coring. After pulling a bit and discovering center coring wear, it is extremely important to break in the next bit carefully to drill the bottom of the hole where the missing cutters left it undrilled (see page 105). Failure to break in the next bit properly could cause tracking problems as well as damage to the new bit. (*Tracking* occurs when the pattern made by the bit on the bottom of the hole matches the pattern of the teeth. The bit then meshes like a gear with the formation and drills very little. See page 47.)

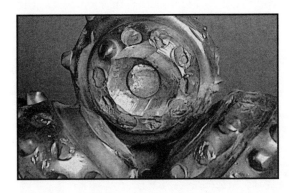

Figure 38. Center coring

To summarize—
Types of cone wear—
- Cone skidding or dragging
- Cone interference
- Cone erosion
- Cracked cones
- Broken cones
- Off-center wear
- Center coring
- Tracking

Cutter Wear

Most forms of cutter wear affect both teeth and inserts. The same type of wear, however, may mean something different for each.

Tungsten carbide inserts usually show little wear after normal drilling. Most damage to insert bits is a result of improper drilling practices or abnormal conditions in the hole or formation. *Broken inserts*, the most common damage to insert bits, are a normal wear characteristic in some formations (fig. 39). If, however, the bit run was unusually short—that is, if the driller had to pull the bit from the hole much sooner than expected because the penetration rate was too slow—then junk in the hole may have broken the inserts, or the driller may have put too much weight on the bit or rotated it too fast. Broken inserts are also caused by improper break-in of the bit, and by striking a ledge in the hole or slamming the bit into the bottom.

What is more, inserts can break if the operator uses a bit that is not designed for a hard formation. The best solution, of course, is to select a bit designed for the formation, but circumstances can make this unfeasible. For example, an insert bit can last so long in a formation it was designed to drill that the operator may keep drilling with it even though it encounters a harder formation lying below the formation the bit was selected for. In such cases, the driller can increase the weight on the bit and decrease the rotary speed while drilling the harder formation. While the rate of penetration may suffer, increased WOB and decreased rpm should prevent the harder formation from breaking the bit's inserts.

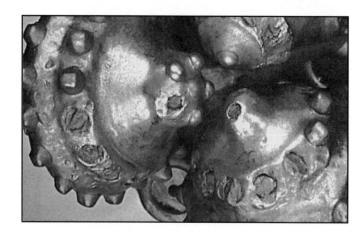

Figure 39. Broken inserts

Inserts can also break because of the high rotary speed often used in harder formations. In this situation, inserts in the gauge row usually break. If gauge row breakage occurs when using a soft-formation bit, change the bit. If gauge row breakage occurs when using a hard-formation bit, reduce the rotary speed.

Broken teeth, unlike broken inserts, are not normal for steel-tooth bits (fig. 40). Broken teeth on a steel-tooth bit indicate that the driller used the wrong bit for the formation or used improper drilling practices, as described earlier for broken inserts.

Chipped cutters are fairly common and do not always indicate a problem. A cutter is chipped, as opposed to broken, when a substantial part of it remains above the cone (fig. 41). Cone interference and rough drilling conditions cause chipped cutters.

Figure 40. Broken teeth

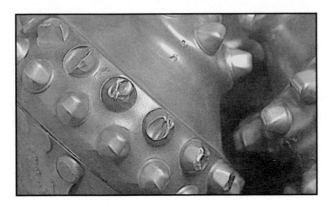

Figure 41. Chipped cutters

Insert loss is caused by cone erosion, cracking, or corrosion. In air or gas drilling, cone erosion can be a serious problem because the jets of air or gas leaving the bit strike the cones with great velocity in a dusty, gritty environment. One of the frustrations of drilling with tungsten carbide insert bits is that, once the inserts start breaking or falling out, the broken pieces on the bottom of the hole are almost certain to damage the bit further because they are so hard. Tungsten carbide is so dense (and therefore heavy) that it is almost impossible to circulate the broken pieces out of the hole. One solution is to run in a junk sub to fish out the inserts after removing the damaged bit. (See Unit III, Lesson 2, *Open-Hole Fishing*, for more information about retrieving junk from the hole.)

Two kinds of normal abrasion wear to teeth are flat-crested wear and self-sharpening wear. Flat-crested wear occurs when the tops of the teeth wear evenly and flat over the bit (fig. 42). Self-sharpening wear is expected and desirable (fig. 43). In bits with self-sharpening teeth, the drilling mud abrades the softer metal on one side of each tooth. The harder metal on the other side does not abrade and so holds a sharp edge. If abrasion wear is excessive, however, the drilling fluid may be too abrasive, or the fluid velocity may be too high.

Bradding is a condition in which the weight on a tooth has been so great that the hardfacing (the layer of tungsten carbide) on the tooth cracks and exposes the softer metal beneath. The softer metal can then fold over the harder part and cover it. Bradding can also cause the teeth to break off. This condition, which affects the inner rows of teeth, usually develops early in the run because of too much weight on bit or excessive rotary speed.

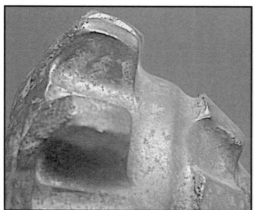

Figure 42. Flat-crested wear

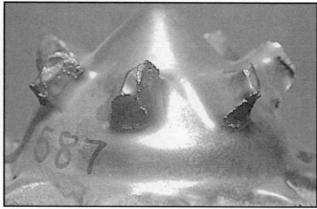

Figure 43. Self-sharpening wear

Tracking is a rare type of tooth and cone wear (fig. 44). As mentioned earlier, it occurs when the pattern made by the bit on the bottom of the hole matches the pattern of the teeth. The bit then meshes like a gear with the formation and drills very little. Tracking can happen when the formation is plastic—that is, it bends under stress without breaking; the opposite of brittle. Using a bit designed for a softer formation may remedy this problem.

When the cutters drag on the formation, they heat up. The drilling fluid then cools them. When this cycle occurs many times, *heat checking* can result (fig. 45). Checking means to develop small cracks.

Gauge rounding, when the gauge cutters wear down in a rounded fashion, may be serious (fig. 46). If the bit can still drill a full-gauge hole, however, the wear is not a problem. Minor gauge rounding can happen when drilling an abrasive formation at too high a rotary speed or reaming an undergauge hole. If, however, the gauge cutters are so worn away that the bit drills an undergauge hole, a problem exists. The next bit will have to ream the hole to the appropriate size, which may damage the new bit and waste time.

In extreme cases of gauge wear, the shirttail and gauge area of the bit are so worn away that the bearings fall out of the bit. Extreme gauge wear can occur when using a bit that is not strong enough for the type of formation being drilled, running a bit for too long, or using a rotary speed that is too fast for the specific bit.

Figure 44. Tracking wear

Figure 45. Heat checking

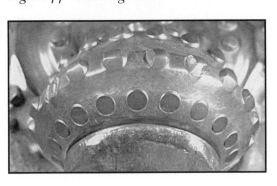

Figure 46. Gauge rounding

To summarize—

Types of cutter wear—
- Broken inserts
- Gauge row breakage
- Broken teeth
- Chipped cutter
- Lost inserts
- Bradding
- Heat checking
- Gauge wear

Bearing Wear

Bearing wear and failure often make it necessary to stop drilling and replace the bit. Spalling and galling used to be the biggest problems with bearings. The advent of sealed bearings has substantially reduced such wear, however. Even so, better design has not eliminated bearing wear altogether.

The *outer bearings* (the roller bearing or journal bearing under the gauge cutters) can fail for any of several reasons. Excessive rotation time coupled with heavy weight can wear them out. Abrasive materials in the drilling fluid can damage nonsealed bearings, as can corrosion caused by sulfur in the drilling fluid. Signs of outer bearing failure are skid marks on the cones and one or more locked cones.

The *inner bearings* (the inside roller bearing and nose bearing under the nose of the cone) can fail when a bit is drilling a formation that is too hard for it. In this case, the teeth wear down to a flat top (they suffer flat-crested wear) and, as a result, the ROP slows. The driller then adds more weight to keep up the rate of penetration. The extra weight can stress the bearings beyond their design limitations.

Another cause of inner bearing wear or failure may be gauge rounding. The stronger gauge cutters wear away so much that the forces of weight and rotation on the bit shift to the weaker inner cone teeth and bearings. Because the inner bearings are smaller and not as strong as the outer bearings, they cannot withstand the extra force.

Reaming out an undergauge hole also stresses the inner bearings. The driller can partially or completely eliminate inner bearing failure by using a bit designed for harder formations, reducing the time the bit is in the hole, decreasing the rotary speed, or choosing a special bit for reaming.

Both the inner and the outer bearings can fail if the weight on bit is too high, the rotary speed is too fast, the bit is used to ream under-gauge hole, an abrasive drilling fluid is in use, or the bit is used too long.

Sometimes a nozzle can fall out of the bit because of improper installation or design or because of damage to the nozzle or its retaining system. The missing nozzle causes a decrease in the hydraulic pressure and becomes junk in the hole. The junk nozzle can damage any part of the body of the bit as well as the cutters (fig. 47).

Any place on the bit that has a weld can be washed out, or eroded, by the drilling fluid if it cracks or is not completely closed (fig. 48). The drilling fluid washes through the crack and erodes it very quickly. Extended nozzles can occasionally break or wash out because the body of the bit does not protect them.

The dull condition of the steel-tooth bit in figure 49 is appropriate for a soft-formation bit. It has a considerable amount of life remaining. There is little tooth breakage or chipping. The tungsten carbide hardfacing has allowed the teeth to wear so that they remain relatively sharp, even though appreciably worn down in height.

Figure 47. Bit damaged by junk

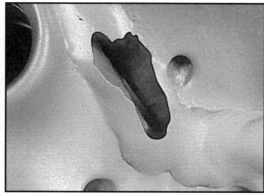

Figure 48. Washout

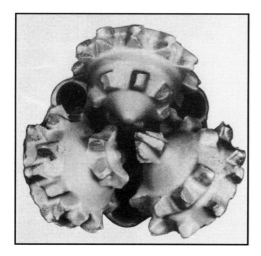

Figure 49. Normally worn bit

49

To summarize—

Causes of bearing failure—
- Excessive rotation time coupled with heavy weight
- Abrasive materials in the drilling fluid in unsealed bits
- Sulfur compounds in the drilling fluid in unsealed bits

Bearing failure indicators—
- Skid marks on the cones
- One or more locked cones

Types of inner bearing wear—
- Flat-crested wear
- Gauge wear

Causes of inner and main bearing failure—
- Too much weight
- Too high a rotary speed
- Reaming an undergauge hole
- Abrasive drilling fluid used with unsealed bits
- Bit kept in service beyond its useful life

Other types of bearing wear—
- Damage caused by nozzles falling out of the bit
- Washouts at welded points on the bit

Diamond Bits

Diamonds are the hardest mineral on earth, almost nine times harder than the next hardest naturally occurring mineral. A diamond will outwear or cut anything without affecting the diamond in any way, with only two limitations: it will disintegrate if the temperature is too high, and it is brittle and is therefore susceptible to fracturing under shock loads.

It is no surprise that tools set with whole or crushed diamonds are common in many industries where a hard material needs to be cut, ground, or polished with precision. For example, tools used in manufacturing cars, airplanes, and space vehicles, cutting glass in the optical industry, or for surgical instruments may use diamonds. Bits for oilwell drilling with natural diamond cutters date from the 1940s.

Diamond tools might be common in every home workshop except for diamonds' rarity and the expense of mining them. As early as 1797, therefore, scientists began trying to create, or synthesize, diamonds from carbon (diamond is pure carbon, as are coal, ash from burning bones or wood, and soft graphite used in pencil leads). Research scientists at General Electric were the first to succeed, in 1954. Synthetic diamonds turned out to be even more expensive to manufacture than natural diamonds are to mine, but as long as the demand exists, manufacturers can produce them because carbon is plentiful.

The synthetic diamonds used now in bits are one of two types: the *polycrystalline diamond compact* (PDC) developed in the early 1970s, and the *thermally stable polycrystalline* (TSP), first made in the late 1970s. Manufacturers produce bits with natural diamonds, PDCs, TSPs, or combinations of these as the cutters, replacing the teeth or inserts of roller cone bits.

The Properties of Diamonds

Diamond is the hardest material known. Nothing can scratch a diamond and a diamond can scratch anything without dulling—the softer the other material, the deeper the diamond can scratch it or, in other words, cut or grind it. Diamond is 2½ times harder than tungsten carbide, a hard manufactured substance, and so has better wear resistance. Diamond also has the best compressive strength and modulus of elasticity of any material—over twice those of tungsten carbide.

Because a diamond has high compressive strength and modulus of elasticity, if you gradually apply a force, or load, to a diamond, it will not break until the load is very high. Instead, it will deform and then recover its shape and size after you remove the force. It is like a foam cushion that compresses when you sit on it and then springs back to its original shape when you stand. Diamond can overcome the strength of any other material in this way.

Diamond also has the best thermal (heat) conductivity of any material. Good thermal conductivity means that heat flows through a substance very quickly. Think of how cold the aluminum frame of a window is in winter. It is cold because the heat from inside the house passes through the thermally conductive aluminum quickly. The colder the outside air is, the colder the aluminum is. In the same way, a diamond can transmit the heat generated by drilling to the drilling fluid as fast as the fluid can absorb it, so the diamond is as cool as the fluid.

Another valuable property of diamond is its low friction force. A diamond under a heavy weight (like the weight placed on a bit) can move against other materials (like formation rocks) without using a great deal of force (as in rotating a bit) and without generating a great deal of heat.

Because of diamond's unique qualities, you might think that diamond bits would drill the hardest formations without ever wearing out. But they do wear out. Two other properties of diamonds—low impact strength and low thermal stability—explain their limits for drilling. Although diamonds do not break under gradual loads and do not scratch, they do break very easily under impact, or a sudden blow. In fact, you could shatter a diamond into dust by just hitting it with a hammer. For this reason, the bit designer devises methods to set the diamonds into the bit to protect them from impact. The operator must also choose the right design for the formation. Most important, the drilling crew must handle and use the bit correctly.

Thermal stability is the ability to withstand high temperatures. Although diamonds are stable up to about 950° Fahrenheit (F) or 495° Celsius (C), temperatures in the borehole can get much higher. When the temperature reaches about 2,350° F (1,270° C), the diamond changes into graphite, one of the softest minerals.

At high temperatures, the carbon atoms of the diamond also begin to react with the atoms of the rock if oxygen is present. It is the same type of chemical reaction that causes iron to rust and wood to burn. The diamond actually wears away, not because of a mechanical process that takes off pieces of diamond, but because of a chemical reaction that removes the individual atoms. Then it is worthless for drilling rock, of course, and an expensive tool is ruined.

Temperatures high enough to damage diamonds are easy to reach in drilling. The thermal instability of diamond is the reason that designers and operators are always trying to improve the hydraulics of bits. They vary the volume, velocity, and flow pattern of the circulating fluid to take advantage of diamond's thermal conductivity to cool the bit as much and as fast as possible.

Diamond bits originally replaced roller cone bits for the hardest, most abrasive formations because diamonds are so much tougher and last longer than even tungsten carbide. During the past 15 years, the number of designs for diamond bits has grown tremendously. Bits using natural diamonds date from the 1940s, and synthetic diamonds came into use in the late 1950s. Now bits may have natural or synthetic diamonds or combinations of both to drill almost any type of formation.

Natural Diamond Bits

Many millions of years ago, diamonds formed in igneous (molten) rock 250 miles (over 400 kilometres) deep in the earth at temperatures of 5,000° F (2,750° C), and under pressures of 1½ million pounds per square inch (psi), or over 10 million kilopascals. Only one diamond deposit in the world discovered so far occurs in an igneous formation, however. Geologists believe that volcanoes carried most diamond-containing rocks to the surface in the lava, which formed cones. As the cones eroded, the diamond-containing rock usually washed into rivers and oceans, where it combined with dead plants and animals to form sedimentary rocks.

Natural diamonds are mined from such sedimentary deposits all over the world: in West Africa, South Africa, Russia, Brazil, Venezuela, China, and Australia. Most diamonds are not of gem quality—that is, they are not transparent and are either colorless or deeply colored. So-called industrial diamonds are usually brown or gray and cloudy or opaque. The diamonds that become cutters in bits weigh from 1/15 to 1 carat (a 1-carat diamond is about ¼ inch in diameter) and are one of several grades, depending on their shape and chemistry.

Industrial diamonds may be rounded, cube-shaped, octahedral (eight-sided), dodecahedral (twelve-sided), or irregularly shaped. They can be monocrystalline (a single crystal) or polycrystalline (many crystals bonded together). The different types are sharper or duller and have varied resistance to abrasion and impact. The designer chooses the grade and size that will work best in different formations or bit designs or even for different locations on one bit.

Because of the greater hardness and lower impact strength of diamond cutters compared with steel teeth or tungsten carbide inserts, the design of a diamond bit is quite different from a roller cone bit. A diamond bit is a *fixed-head*, or *shear*, *bit*. Instead of having three independently moving cones, a diamond bit has a stationary (a fixed) head that rotates as one piece with the drill string. It is a shear bit because it cuts the rock by slicing it like a knife, instead of gouging and crushing the rock, as a roller cone bit does. A diamond bit has three main parts: the cutters, the body, and the shank (fig. 50).

Natural Diamond Bit
Manufacture and Design

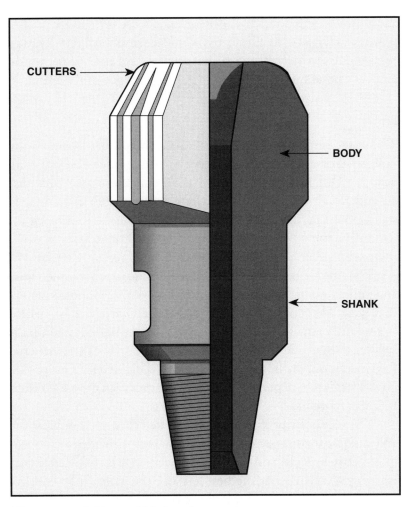

Figure 50. A diamond bit has three main parts.

The cutters are the diamonds arranged in rows on the *nose* (the part of the bit that touches the bottom of the hole) and the side (*flank* and gauge area) of the bit. The body is the main section and holds the diamonds. The shank is a steel base for the body that gives structural strength and provides a place for the threads to make up the bit on the drill string. The body of the bit is made of *matrix*: a mixture of tungsten carbide powder and a bonding metal to help hold it together. Matrix wears better than steel because it is harder.

To manufacture the bit, a technician carefully places the diamonds by hand into the bottom of a mold. Matrix powder mixture, which is very dense, goes in next, the steel shank rests on the powder, and then the mold is filled up with more matrix powder. A furnace then heats the whole assembly at a temperature high enough to melt the matrix but not the steel and the diamonds, and they bond into one piece.

Profile

The shape of the body of the bit, called the profile, is one of the design factors that defines what type of formation the bit will work best in. Different profiles have different advantages and disadvantages. Profiles range from a long flank (or side) and relatively sharp nose to a flat nose and short flank. Four basic profiles for natural diamond bits are single-cone, double-cone (do not confuse these with roller cones), parabolic, and concave.

The single-cone profile (also called round) has a rounded nose and puts a lighter load on each cutter for even wear and a long life (fig. 51). The double-cone profile has a sharper nose, which aggressively attacks the formation to make an initial hole that is smaller than the final hole diameter will be (fig. 52). Its tapered flank reams out the hole to gauge (the final diameter). This profile gives a faster rate of penetration, but its cutters can wear out faster. It is also more likely to stay centered in the hole.

The parabolic profile is more rounded than the double-cone and sharper than the single-cone (fig. 53). It provides a compromise between ROP and wear resistance—that is, both ROP and wear resistance are not the best and not the worst. The concave profile is very flat and short (fig. 54). This type of bit can drill directional or horizontal holes.

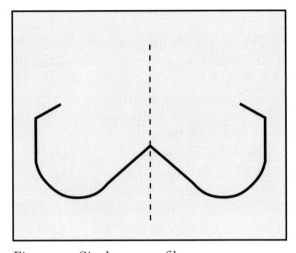

Figure 51. Single-cone profile

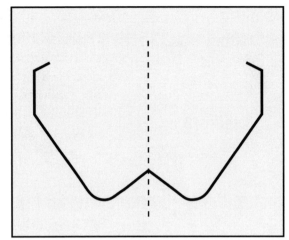

Figure 52. Double-cone profile

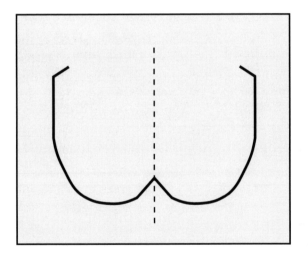

Figure 53. Parabolic profile

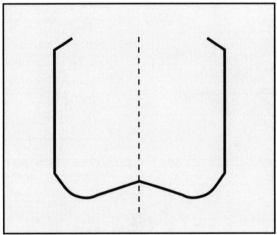

Figure 54. Concave profile

When deciding which cone profile to use, the designer also considers how much surface area is available for the diamonds. A single-cone profile, for example, has more space for diamonds than a double-cone: the more diamonds there are, the lighter the load each one must bear.

Cutters

The bit will last longer if the load on the cutters is distributed evenly from the nose to the gauge area. The goal is to place the cutters so that each cutter removes the same amount of rock each time the bit rotates. To accomplish this goal, designers, with the aid of computers, have devised various *plots*, or patterns, in which to set the diamonds.

In the *grid plot*, the diamonds are evenly spread over the cutting surface of the bit (fig. 55). This plot makes cleaning the cutters easy and therefore gives a high rate of penetration in soft formations.

In the *circle plot*, the diamonds sit in concentric circles (circles of different sizes that have the same center) (fig. 56). It uses more diamonds over the same area than the grid plot, and so the bit lasts longer because each diamond has to withstand less load.

Designers use various sizes and grades of natural diamonds for different purposes. Smaller diamonds, eight to ten diamonds per carat, are often *ridge set* (fig. 57). The diamonds sit inside raised ridges made of matrix and are flush with the surface of these ridges. The matrix surrounds the diamonds and protects them from breaking from the impact shocks of drilling harder and more abrasive formations. Larger diamonds, up to 1 carat, are *surface set*. In this arrangement, about two-thirds of the diamond is buried in the matrix of the bit body and one-third is exposed above it. Embedding the diamonds deeply into the matrix protects the diamonds from impact.

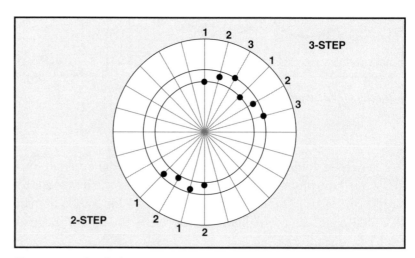

Figure 55. Grid plot

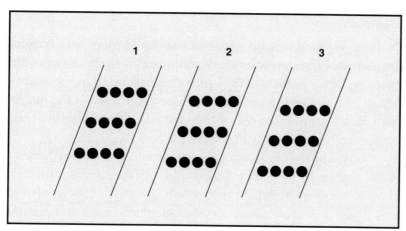

Figure 56. Circle plot

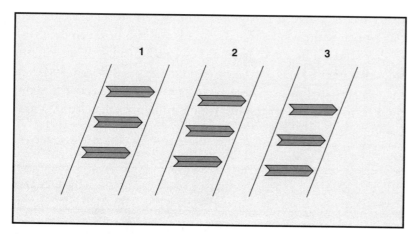

Figure 57. Ridge plot

Hydraulics

Natural diamond bits do not use jet nozzles, as roller cone bits do. Instead, they have an outlet for the drilling fluid in the center of the bit. The outlet leads the fluid into channels cut into the matrix body. This causes the fluid to travel across the face of the bit. As with rock bits, the purpose of the hydraulic design is to cool the cutters and clean the cuttings from the hole.

The channels for directing the fluid come in two main designs: *radial flow* and *cross-pad flow* (also called *feeder-collector*). In both types, the center outlet through which the fluid flows is called a *crow's foot*. The crow's foot has three channels (like a bird's toes) that cause the fluid to split from a single stream into three streams.

In a radial-flow bit, the fluid flows from these crow's foot openings into watercourses that radiate out from the center directly to the gauge area (fig. 58). The watercourses are channels that sit below the level of the diamonds and run between each row of diamonds. They may have either parallel sides or expanding sides that get wider as they approach the gauge of the bit, and they may also get shallower. These variations control the velocity of the fluid across the face of the bit, which influences the cleaning and cooling power.

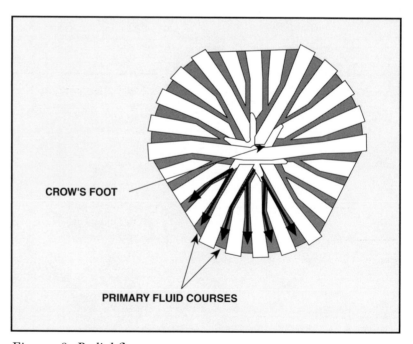

CROW'S FOOT

PRIMARY FLUID COURSES

Figure 58. Radial flow watercourses

The feeder-collector, or cross-pad, type of watercourse has channels similar to radial-flow channels; these channels are called feeders. It also has a second kind of channel called collectors. The collectors do not connect directly to the feeders or to the crow's foot. They are V-shaped and start close to the gauge area and quickly widen (fig. 59). This arrangement creates a low pressure in the collectors that sucks the fluid from the feeders across the top of the diamonds. Fluid moving directly over the diamonds (cross-pad flow) cools them better than fluid moving beside them in the watercourses (radial flow).

The radial-flow hydraulic design removes the greatest quantity of cuttings, which is important when the rate of penetration is faster. The feeder-collector pattern cools better and so it works best in harder and more abrasive formations. Stone size also affects what type of hydraulics the bit will need. Larger stones are easier to clean and cool because of the greater distance between the matrix and the top of the diamond.

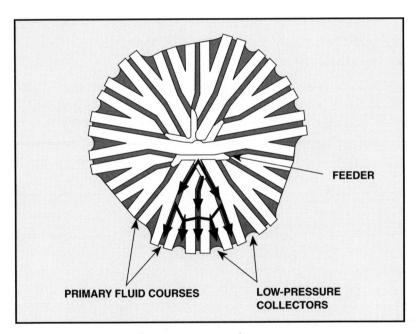

Figure 59. Feeder-collector, or cross-pad, watercourses

To summarize—

Natural diamond bit designs—

- Single-cone
- Double-cone
- Parabolic
- Concave

Diamond bit plots—

- Grid plot—the manufacturer evenly spreads the diamonds over the bit's cutting surface
- Circle plot—the manufacturer sets the diamonds in concentric circles

Diamond settings—

- Ridge set—diamonds sit inside raised ridges in the matrix and are flush with the ridges
- Surface set—the manufacturer buries two-thirds of the diamond in the matrix of the bit body and leaves one-third of the diamond exposed above the matrix

Channel designs—

- Radial flow—the fluid flows into watercourses that radiate from the center of the bit directly to the gauge area and alongside the diamonds
- Cross-pad flow—the fluid flows through radial channels and collectors, which move the flow across the diamonds

Synthetic Diamond Bits

PDC Bits

The polycrystalline diamond compact, or PDC, is one of the most important advances in recent years. Since their first production in 1976, bits using PDC cutters have become as common as roller cone bits. Each polycrystalline diamond compact, which is the cutter, is a special type of synthetic diamond bonded to tungsten carbide (fig. 60).

To manufacture a PDC, a special furnace heats carbon to a very high temperature and, under great pressure, forms many tiny (about 0.00004 of an inch, or 0.00102 of a millimetre) diamond crystals ("poly" means "many"). The manufacturer then mixes the diamonds with a metal powder, called the catalyst metal, and puts the mixture into a can. A layer of tungsten carbide powder mixed with diamond is added, and, finally, a polycrystalline wafer is put on the top.

The manufacturer then seals the can and heats it under high pressure. The pressure and heat cause the diamonds to bond with each other and with the tungsten carbide to make a polycrystalline diamond compact (a compact is any object produced by compressing metal powders). The tiny diamond crystals in a PDC face in different directions, which makes the PDC very strong, sharp, and wear-resistant. It is self-sharpening because when a layer of crystals wears away, another layer with its many sharp edges is exposed.

The main disadvantage of PDCs is that they are even less stable at high temperatures than are natural diamonds. As the PDC contacts the rock during drilling and heats up, the catalyst metal expands at a different rate from the diamond. Eventually, when the temperature reaches about 1,350° F (750° C), the PDC cracks. This temperature is not difficult to reach in drilling and is much lower than the temperature at which a natural diamond disintegrates (2,350° F, or 1,270° C).

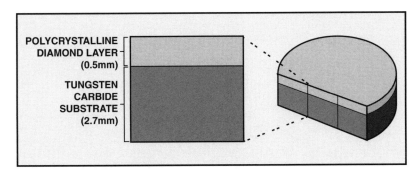

POLYCRYSTALLINE
DIAMOND LAYER
(0.5mm)

TUNGSTEN
CARBIDE
SUBSTRATE
(2.7mm)

Figure 60. A polycrystalline diamond compact (PDC)

PDC Bit Manufacture and Design

Although bits with PDC cutters look different from natural dia-mond bits, they are still fixed-head bits—that is, they rotate as one piece. The body of the bit can be either steel or matrix.

A steel-body bit starts as a length of solid steel that the manu-facturer turns on a lathe to create the threads for attachment to the drill string. The manufacturer then machines the watercourses, nozzle ports, and pockets for the PDC cutters and may add a tungsten carbide hardfacing. The cutters are brazed (a type of soldering) into the pockets.

A steel body is easier to manufacture to exact dimensions and is less brittle than matrix. Matrix, however, wears better than steel because it is harder. Steel-body bits may have a tungsten carbide hardfacing, as roller cone bits do, and they are repairable. The manufacturer can replace the cutters, build up areas of light erosion, reapply the hardfacing, and restore the threads.

Matrix-body PDCs work best when erosion of the body is likely to cause the bit to fail, such as in smaller holes where the drilling mud flows at a high velocity or when the mud contains a high proportion of solid material (in the same way that a fast-flowing or sandy river erodes a rock to make it round).

Profiles

The PDC bit may have one of three basic profiles: short parabolic, shallow-cone, or parabolic. The *short parabolic* is a rounded, general-purpose profile that provides even wear on the cutters (fig. 61). The bit life is good at fast rotary speeds in larger holes and its smaller surface area makes it easy to clean.

The *shallow-cone* profile has the smallest surface area for the easiest cleaning (fig. 62). It works well in smaller holes and in formations with hard *stringers* (narrow splinters of rock different from the main formation rock).

The *parabolic profile* is best for high rotary speeds and with a downhole motor, but it requires a high velocity of drilling fluid to clean it (fig. 63). A *downhole motor* is a device, such as a turbodrill, that crew members make up in the drill string. When they attach a bit to the downhole motor and start the mud pumps, the mud flowing through the motor causes the attached bit to rotate. When using a downhole motor, the driller does not have to rotate the drill string to rotate the bit. Not having to rotate the drill string to drill is advantageous when drilling directional or horizontal wells. Crew members can install devices with an angle to deflect the hole from vertical. One common device is a bent sub.

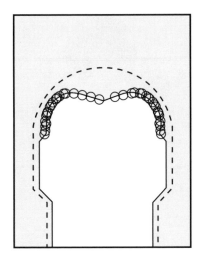

Figure 61. Short parabolic profile

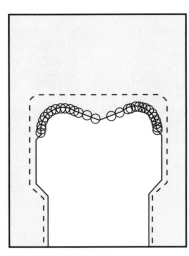

Figure 62. Shallow-cone profile

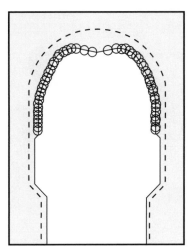

Figure 63. Parabolic profile

Cutters

PDC cutters range in size from about ⅜ inch to 2 inches (9 to 50 millimetres), but they are most commonly about ½ to ¾ inch (13 to 19 millimetres) in diameter (fig. 64).

The most common shapes of the PDC cutter are the cylinder (its shape when it comes out of the manufacturing can) and the stud, which is a cylinder bonded to a tungsten carbide post (fig. 65). The PDC layer on top of the cylinder can be flat or domed. The edge that meets the formation may be rounded to help prevent the cutter from chipping.

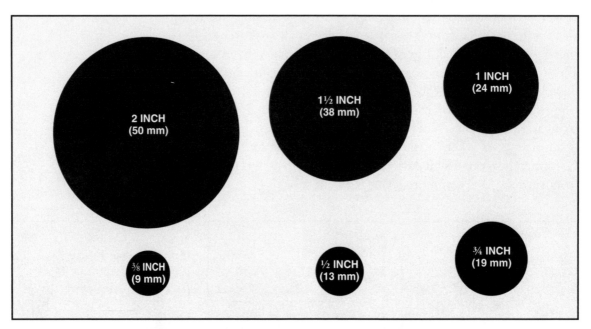

Figure 64. PDC cutters range in size from ⅜ inch to 2 inches (9 to 50 millimetres).

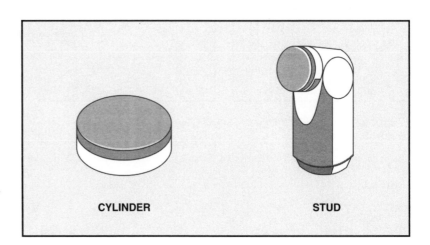

Figure 65. Two PDC cutter shapes are the cylinder and the stud.

The stud protrudes from the body of the bit. The extra support that the tungsten carbide post gives the cutter makes the stud design stronger than the cylinder cutter, and more of the cutter actually digs into the formation.

Because the PDC cannot withstand the heat used to cast the matrix body of the bit, the cylinder cutter is brazed into a pocket in the bit after casting. Stud cutters are essentially already in their own pockets (the stud) and the whole piece is brazed or press-fitted (snapped into a tight cavity) onto the bit.

The cutters sit in the bit at certain angles called *rake angles*. The *back rake angle* is how far the cutter leans back from the vertical, away from the rock it is cutting (fig. 66). A cutter that is perpendicular to the formation has no back rake angle. A cutter that tilts forward into the rock has a positive rake, but this angle is not practical.

The amount of back rake determines the aggressiveness of the cutter—no back rake is the most aggressive. Harder formations require more back rake to prevent the cutter from breaking and to reduce chatter, or vibration. Even on the same bit, cutters at different locations may have different back rake angles to help distribute the load among the cutters. Back rake also affects how much weight on the bit is practical—the greater the back rake angle, the less the WOB influences the action of the bit and so the more uses the bit has under different conditions.

The *side rake angle* is the left-to-right orientation of the cutter with respect to the bit's face (fig. 67). The designer angles the cutters toward the outside of the bit to help direct the cuttings toward the annulus, so that the cuttings do not pass in front of the cutter again to be reground. The optimum side rake angle improves the cleaning and drilling efficiency of the bit. It also helps to stabilize the bit to keep it from vibrating.

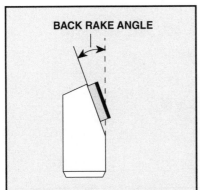
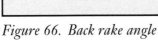
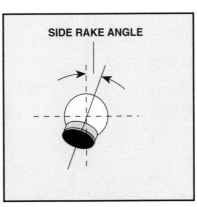

Figure 66. Back rake angle   Figure 67. Side rake angle

The most obvious design difference between natural diamond bits and PDC bits is how the cutters fit into the bit. PDC cutters sit sideways inside pockets, either individually or on blades that are part of the body of the bit (fig. 68). In this way, they look more like one cone of a steel-tooth roller cone bit than like a natural diamond bit.

Bits with the PDCs arranged individually all over the surface are sometimes called porcupine bits. Bladed PDC bits can have from three to more than twenty blades, each with many or only a few cutters. The more blades and cutters, the more durable the bit is, but the slower the penetration rate.

Hydraulics

The design of the hydraulic system is more critical in PDC bits than in other types. PDCs cut deeper than natural diamonds because they are larger, which produces more cuttings to clear away. The cooling function of the drilling fluid is also crucial because of a PDC's thermal weakness. The harder the formation, the more important the cooling function is to prevent the cutters from disintegrating. The softer the formation, the more important the cleaning function is to prevent balling.

Like roller cone bits, PDC bits use jet nozzles as outlets for the drilling mud (fig. 69). The nozzles come in various sizes and are usually interchangeable (but not with roller cone bit nozzles). The nozzles direct the cuttings into *junk slots* on the outside diameter of the bit. The junk slots work like the collectors on natural diamond bits—that is, they provide a low-pressure area that pulls the fluid with the cuttings up and away from the nose of the bit.

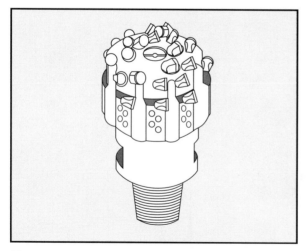

Figure 68. PDC cutters in the body of the bit

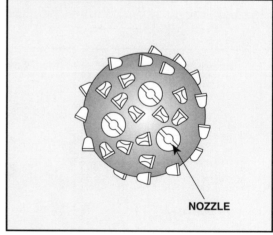

Figure 69. Jet nozzles on a PDC bit

Each nozzle usually points toward the bottom of the hole or at a slight angle and cleans and cools a group of cutters or, on some bits, a single cutter. The number of nozzles depends on the design of the rest of the bit. For instance, the designer can put fewer nozzles on a bit with fewer blades because the cutters are more exposed and the fluid can reach them more easily.

The newest synthetic diamond for bits is the thermally stable polycrystalline diamond, or TSP. Its method of manufacture is similar to that of the PDC. The difference is that the manufacturer either leaches out the catalyst metal that in a PDC helps bond the crystals or replaces it with a less-temperature-sensitive material.

TSP Bits

A TSP is stable at higher temperatures than a PDC (almost as stable as natural stones) because the crystals do not break apart because of expansion. Where a PDC's diamond layer will disintegrate at 1,350° F (750° C), the TSP can withstand temperatures up to 2,200° F (1,200° C). The TSP, like the PDC, stays sharp as the tiny diamond crystals wear away.

A TSP bit is more like a natural diamond bit than a PDC bit is because a TSP is much smaller than a PDC and closer in size to the natural diamonds used in bits (only ⅓ carat, about 1/12 inch, or 2 milli-metres, in diameter). The TSP may be round or triangular (fig. 70). It usually functions on the bit as a cutter by itself, like a natural dia-mond, and is cast into the matrix, rather than being brazed into pockets, as PDC cutters are. It does, however, sit sideways in the bit like a PDC.

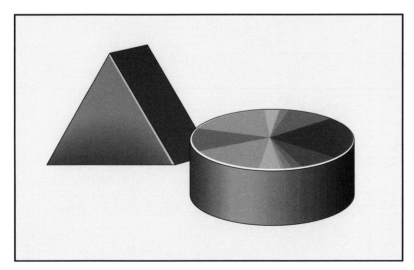

Figure 70. TSPs are triangular or round in shape.

The bit itself is a fixed-head bit and can have the same profiles as a natural diamond bit—single-cone, double-cone, and parabolic. The hydraulic system of a TSP bit is also like that of the natural diamond bit. It has no nozzles; the drilling fluid comes out of the center opening and passes over the cutters in a radial flow or feeder-collector pattern. The TSP bit has a slower rate of penetration than a PDC, again like the natural diamond bit, because of its small cutter size.

To summarize—

PDC profiles—

- Short parabolic—rounded, general-purpose profile that provides even wear on the cutters
- Shallow-cone—small surface area for easy cleaning in small holes and in formations with hard stringers
- Parabolic profile—like the short parabolic only longer; best for high rotary speeds with a downhole motor

PDC cutter angles—

- Back rake angle—how far the cutter leans back from vertical
- Side rake angle—the left-to-right orientation of the cutter with respect to the bit's face

TSP bits—

- are more stable at high temperature
- are round or triangular
- are fixed-head bits
- do not have nozzles

Designers have taken advantage of the unique properties of each of the various materials used to make bit cutters and have combined them on one bit. Called *hybrid bits*, they combine natural stones, PDC cutters, TSPs, and even tungsten carbide inserts. For example, some PDC bits use natural diamonds, TSPs, or tungsten carbide inserts as the gauge cutters.

Manufacturers of hybrid bits sometimes place a diamond-impregnated pad or stud (not the same as a PDC stud cutter) on the bit for gauge protection. To make such pads or studs, the manufacturer mixes grit-sized natural diamonds and tungsten carbide powder and heats them under pressure to form a replaceable stud or a pad to attach to the gauge surface (fig. 71).

Using the more thermally stable types of diamond (TSPs) in the gauge area prolongs the life of the bit and allows it to remain on the bottom longer. The slower rate of penetration of TSPs is unimportant on the gauge area because the TSPs are only reaming or maintaining the hole size, not doing the initial drilling.

Hybrid Bits

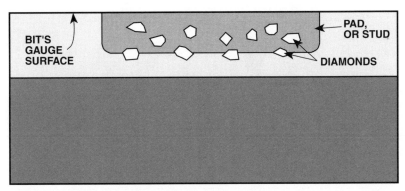

Figure 71. Diamond-impregnated pad, or stud, on bit gauge surface

Another combination puts a shorter, diamond-impregnated backup stud behind the PDC cutter (fig. 72). When the bit is new, the PDC cutter does all the work, but when the PDC cutter starts to wear down, the backing stud contacts the formation and takes part of the load (and therefore the heat) off the PDC. This design works well for formations that are softer at first, but then become harder. It also works well in soft formations that have hard stringers. First, the new PDC penetrates the soft rock quickly, and then it has help drilling the harder rock as it wears down. Manufacturers are now putting diamond-impregnated studs on the gauge cutters of roller cone bits.

Hybrid bits have extended the range of diamond bits to harder and more abrasive formations and are capable of a faster rate of penetration because of these improvements. Undoubtedly, new ways of taking advantage of diamond's unique properties are yet to come.

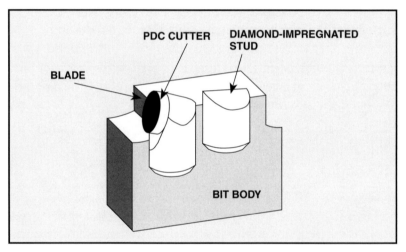

Figure 72. A diamond-impregnated backup stud behind a PDC cutter

To summarize—

Characteristics of hybrid bits—

- Combine natural diamonds, PDCs, TSPs, and tungsten carbide inserts
- Can give high rates of penetration in very hard, abrasive formations
- Can prolong bit life

The designer must solve the problems of where and how to place the diamonds, which type of diamond to use (natural or synthetic), and how to design the rest of the bit so that the diamonds do not break from impact or disintegrate from heat. The driller must then use the bit in the correct formations and under the proper conditions. Designers now use computers to help predict where and how a bit will wear. Some manufacturers can even produce custom bits for a specific drilling operation.

Diamond bits are very expensive, so avoiding preventable wear is important. Some of the same types of damage described for roller cone bits can affect the cutters, the gauge surfaces, and the body of a diamond bit. Some types of damage, however, are unique to diamond bits, because of diamond's low impact strength.

In 1989, a major operating company identified a phenomenon called *bit whirl* as one of the main causes of damage to diamond bits. Bit whirl can be a problem for any bit, but it is especially damaging to PDC bits. The bit whirls when its center of rotation is not in the center of the hole. Instead of rotating in a circular motion, the bit turns with a spiraling action (fig. 73). Because of the bit's spiraling action, the center of rotation at any given instant can be one of the gauge cutters. While whirling, the innermost cutters move backward and the gauge cutters pound against the side of the hole. This pounding breaks the cutters. The bit also drills an overgauge hole when it whirls.

Whirling can have several causes, including poor drilling practices. Once it begins, it is difficult to stop. It tends to happen at high rotary speeds and at low WOB. Also, any time a bit drills slowly, as, for instance, it often does in a very hard formation, whirling can occur. Manufacturers have, however, designed special antiwhirl bits to solve the problem.

Diamond Bit Wear

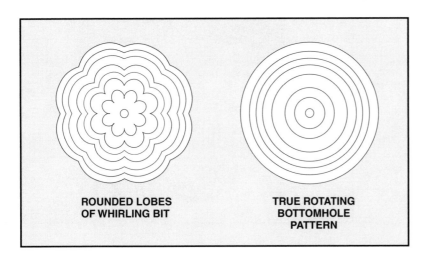

ROUNDED LOBES
OF WHIRLING BIT

TRUE ROTATING
BOTTOMHOLE
PATTERN

Figure 73. Bottomhole pattern caused by bit whirl and normal rotation

Reaming with a diamond bit can cause severe damage to the gauge area of the bit. If an operator starts the hole with a roller cone bit and then switches to a diamond bit, the diamond bit may need to be smaller to avoid reaming. Sometimes the roller cone bit drills a hole that is spiral-shaped (fig. 74). Cones have no problem rolling through these spirals, but a diamond bit with its fixed outside diameter must ream such sections. If reaming is necessary, use a low rotary speed and a light WOB to minimize damage.

It is important to keep the hole clean when drilling with a diamond bit. Any loose junk in the hole flies around and can shatter or chip the diamonds on impact. Be aware, also, that if a roller cone bit drilled the first part of the hole, pieces of its teeth and bearings could still be there. What is more, take care not to let tools fall into the hole and clean out any junk with a junk sub or other fishing tool.

Figure 74. A spiral-shaped hole

PDC cutters can fail in a number of ways (fig. 75). The bond between the tungsten carbide backing and the PDC layer can fail, or *delaminate*. Delamination causes the diamond layer to fall off. This leaves only tungsten carbide to cut the rock. Although carbide is also very hard, the backing is not shaped like a carbide insert cutter and is not meant to be the cutter. The bit will therefore no longer work the way it should.

The bond between the tungsten carbide stud and the tungsten carbide of the compact can also fail, again leaving only tungsten carbide to cut the rock. This is called *long substrate (LS) bond failure*.

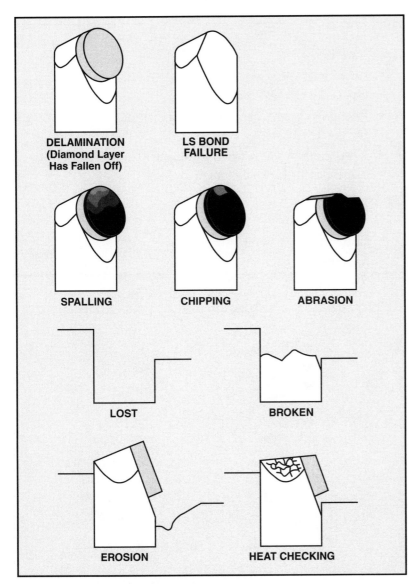

DELAMINATION
(Diamond Layer
Has Fallen Off)   LS BOND
                         FAILURE

SPALLING   CHIPPING   ABRASION

LOST   BROKEN

EROSION   HEAT CHECKING

Figure 75. PDC cutter wear and failures

Layers can flake off the top of the PDC. Such flaking is termed *spalling*. The PDC can chip—impact causes tiny pieces to break off. Chipping does not affect the cutter's performance. If large chunks break off, however, drilling ability is affected.

A PDC can also wear because of *abrasion*—a flat spot appears. Abrasion refers to cutter wear. (*Erosion* refers to wear of the steel or matrix body between the cutters.) Erosion that causes the PDC to fall out is a problem, but some erosion is normal and does not affect the bit's performance. *Heat checking* is cracking of the PDC caused by heat.

To summarize—

- Bit whirl is the main cause of damage to diamond bits, especially to PDC bits.
- Reaming undergauge hole with a diamond bit can damage the gauge cutters.
- Keep the hole clean when drilling with any type of diamond bit. Junk shatters diamonds.

Types of wear—

- Delamination
- Long substrate bond failure
- Spalling
- Abrasion

Special-Purpose Bits

Besides ordinary roller cone and fixed-head bits, manufacturers produce many bits that have features that suit them for a particular purpose. Any roller cone bit may have extra hardfacing on the shirttail, and fixed-head bits may have diamond-impregnated gauge protection for better wear. Extended nozzles are a popular option for soft-formation rock bits when balling may be a problem.

Roller Cone Bits

Air bits circulate air or gas as the drilling fluid. They have screens over the bearings to protect them from clogging with cuttings. They also have thicker hardfacing on the shirttail to protect them from the abrasive, high-velocity air or gas drilling fluid. *Two-cone bits* can sometimes drill very soft formations faster than standard three-cone bits.

Operators sometimes use *jet deflection bits* for directional drilling (drilling a slant hole) in soft formations (fig. 76). Jet deflection bits have an oversize jet nozzle. Without rotating, the driller runs the bit to bottom, points the oversize nozzle in the direction necessary to start the deflected hole (orients it), and starts the mud pump.

Figure 76. A jet deflection bit

Since the driller is not rotating the bit, the oversize nozzle washes out the formation and forms a pocket in the wall of the hole. This pocket helps start the deflection.

Bits for use with downhole motors and turbines have extra tungsten carbide hardfacing on the shirttail, gauge, and teeth to protect them from the high rotary speeds typical when drilling with downhole motors and turbines. Manufacturers also sell diamond bits modified for use with downhole motors and turbines.

Fixed-Head Bits

Antiwhirl bits are the latest advance in fixed-head bits (fig. 77). One solution to whirling is to purposely unbalance the bit so that its center of rotation moves to a low-friction gauge pad. Think of a tire on a car. Its center of rotation is the axle if the car is jacked up and no other forces are acting on the tire. But when it touches the road and starts moving, this center shifts to the outside of the tire, where it meets the road.

Shifting the bit's center of rotation to the special gauge pad forces the bit to press against the wall of the hole, like a tire presses against the road. The gauge pad has tungsten carbide inserts and TSPs to bear the extra load and a covering layer that reduces friction. Antiwhirl bits make hole faster, because they are not wasting energy drilling an overgauge hole. They also allow drilling in harder formations than standard PDC bits allow because of the lower impact on the cutters.

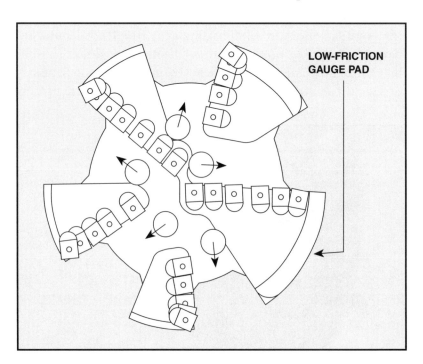

LOW-FRICTION GAUGE PAD

Figure 77. An antiwhirl bit

Eccentric bits drill the hole slightly overgauge (fig. 78). Drillers use them in shales that can swell after being drilled or in salt formations that deform and swell after drilling. Either condition reduces the gauge of the hole, which can lead to stuck pipe and other tight-hole problems. Eccentric bits can also ream out undergauge holes.

Core bits are shaped like a ring (fig. 79). The ring drills the formation on both its inside and outside circumference, so it has two gauge surfaces. The center hole surrounds a solid cylinder of rock (the core) that the driller recovers later. Once the core is retrieved, the operating company sends it to a laboratory for analysis.

Sidetracking bits, when made up on a downhole motor or turbine, drill around broken drill pipe or casing that is permanently stuck in the hole. (Drilling around unremovable objects is a form of directional drilling.) These bits have a flat profile and a short gauge length (fig. 80). Some have large fluid outlets so that a high volume of drilling mud can circulate without losing pressure across the face of the bit.

Figure 78. An eccentric bit

Figure 79. A core bit

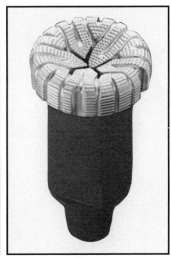

Figure 80. A sidetracking bit

To summarize—

Types of special-purpose bits—

- Air bits
- Two-cone bits
- Jet deflection bits
- Antiwhirl bits
- Eccentric bits
- Core bits
- Sidetracking bits

Bit Performance

The type of formation, weight and rotary speed of the bit, and hydraulics affect the performance of a bit. What a driller gets out of a bit depends on attention to such factors and using good operating procedures.

To get the most footage and fastest penetration rate, and therefore the lowest cost, the operator or contractor must choose the right bit for the job. Operators have several ways of getting information to make this decision. Dull bit records from nearby wells show wear to the bits used to drill them. Another type of record is the driller's log. For each well, the driller keeps a record of the depth, type of rock, fluids, and anything else interesting about the operation. Both records are helpful when drilling subsequent wells in the same field.

To analyze a formation being drilled for the first time, the operator can do two things:

1. Look at a core.
2. Test the rock with the help of computers and sophisticated downhole instruments. For example, sending and recording electrical impulses, sound waves, gamma rays, or neutron rays through a formation identifies the rocks and reveals some of the formation's properties, such as hardness, compressive strength, porosity, and permeability.

Ten properties of rock influence the selection of the bit, WOB, rotary speed, and hydraulics:

1. *Hardness*—how easily the rock is scratched.
2. *Abrasiveness*—how easily the rock wears away the tools used on it because of friction.
3. *Compressive strength*—how easily the rock breaks under a gradual load—that is, when it is compressed. The compressive strength of rock increases with depth and thus it becomes more difficult to break.

Formation Properties

4. *Elasticity*—the ability to return to the original shape and size after being compressed (like a rubber ball). Elastic rocks tend to bounce back instead of breaking, so they are more difficult to drill. Elasticity is related to compressive strength.

5. *Plasticity*—the ability to deform into another shape without breaking after being compressed or subjected to an impact (like modeling clay). Brittleness is the opposite of plasticity. All sedimentary rocks are brittle at the temperature and pressure of the earth's surface, but become plastic under very high pressure.

6. *Porosity*—the ratio of rock grains to pores (spaces between the grains). The higher the percentage of pores, the more fluid (water, oil, or gas) the rock can hold and the easier it is to drill.

7. *Permeability*—how well the pores connect to each other so that the oil or gas can move within the rock.

8. *Pore pressure*—the pressure that the fluids in the pores exert on the rock grains. This pressure combined with the drilling mud pressure influences the ROP because of rock's tendency to become plastic under high pressure.

9. *Overburden pressure*—the pressure that the overlying formations exert; this pressure increases with depth.

10. *Stickiness*—how easily the rock cuttings combine with the drilling fluid to form a mass or ball that sticks together or to the bit.

Also, the way a rock responds to the bit changes as the pressure changes. *Overburden pressure*—the pressure that overlying formations exert on the formations below them—increases as the hole becomes deeper, which changes the volume and weight of the drilling fluid needed to compensate for that pressure. The combined pressure of the formation and the drilling fluid can change the plasticity of the rock and how it breaks up. This in turn changes the requirements for the bit.

Roller cone bits, natural diamond bits, and PDC bits work with different actions to drill the rock. These actions work best under different conditions and they all have their place in drilling operations. Scraping, gouging, plowing, and shearing are all forms of scratching a rock with something harder than the rock—either steel, tungsten carbide, natural diamond, or synthetic diamond.

A roller cone bit scrapes or gouges and crushes the formation with the help of the heavy weight on it. It scrapes or gouges any rock softer than the metal of the bit, and at the same time it compresses or squeezes the rock until it breaks (fig. 81). This action is not very efficient because it requires a great deal of energy for the amount of rock it drills.

A natural diamond bit plows the rock—that is, it pushes the rock aside to form a groove like a plowed furrow (fig. 82). It can also grind the rock, as a millstone grinds wheat into flour. This type of bit drills very slowly, and it is also not very efficient. It works well in harder and more abrasive formations, however, because the diamonds are always harder than the rock.

A PDC bit shears, or slices, the formation (fig. 83). Shearing takes only one-third the energy of crushing, requires less weight on the bit, and drills faster.

How Bits Drill

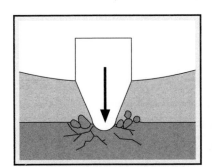

Figure 81. Drilling action of a roller cone bit

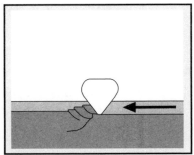

Figure 82. Drilling action of a natural diamond bit

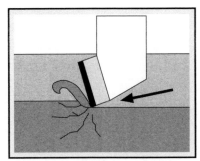

Figure 83. Drilling action of a PDC bit

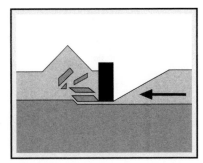

Figure 84. Drilling action of a TSP bit

83

A TSP bit has two actions—it shears a softer formation, but plows a harder one (fig. 84). Shearing and plowing is yet another way this bit combines the characteristics of PDC and natural diamond bits. Either action is more efficient than the crushing action of a roller cone bit. A lower weight on the bit translates to less wear on the bit and drill string and makes it easier for the driller to control bit whirl.

To summarize—

Properties of rock that influence bit selection, WOB, rotary speed, and hydraulics—

- Hardness
- Abrasiveness
- Compressive strength
- Elasticity
- Plasticity
- Porosity
- Permeability
- Pore pressure
- Stickiness
- Overburden pressure

How bits work—

- Roller cone bits scrape or gouge and crush the formation
- Natural diamond bits plow and grind rock
- PDCs shear and slice the formation
- TSPs shear soft formations but plow harder ones

Bit Selection

 oller cone bits can drill all types of rock. Designers originally developed tungsten carbide insert bits to drill extremely hard, abrasive cherts and quartzites that were a problem for steel-tooth bits, but now insert bits are available that can drill any formation.

On both insert bits and steel-tooth bits, the number of rows per cone, the number of teeth or inserts per row, and the length and shape of the cutters help determine the type of formation a roller cone bit will work best in. Two widely spaced rows of cutters or teeth work well in softer formations such as sands and clays.

The teeth on soft-formation steel-tooth bits are the longest of all. The inserts on soft-formation insert bits are long and chisel-shaped, with sharp crests. Both drill large cuttings. The wide spacing allows the intermeshing of the cutters to clean the bit. The high offset of the roller cones causes the scraping action. The cutters penetrate the formation and they drag and twist over it before moving on. As discussed earlier, they work something like a shovel in dirt.

Soft formations may contain abrasive sand that sandblasts, or erodes, the bit cones at the high rotational speeds and mud circulation rates involved in drilling. So the steel of the cone of an insert bit is carburized (hardened) to be more resistant to abrasion. The cone is also thicker so that it does not wear away and cause the inserts to fall out. The teeth of a milled bit may have a tungsten carbide hardfacing for the same reason.

Roller Cone Bits

Bits for medium-soft formations such as shales and soft limestones and sandstones have a design similar to that of soft formation bits. Less offset, however, gives a twisting-scraping action to the cutters on the bottom of the hole. Because of the greater hardness of the formation, the cutters need to be slightly stronger and shorter, and each cone has more cutters that are closer together than on a soft-formation bit. Enough space, however, remains between cutters to allow the drilling fluid to carry away the fairly large cuttings easily. The inserts on insert bits are still chisel-shaped. These formations may also be abrasive, so the teeth and gauge surfaces on milled bits can have tungsten carbide hardfacing. The gauge teeth have the thickest hardfacing because they receive the most wear in medium-soft formations.

Medium-hard formations such as limestone, dolomite, hard shale, chert, and quartzite are too hard and sometimes too abrasive for bits designed to drill medium-soft formations. The cutters on medium-hard formation bits are therefore shorter and do not penetrate very far into such rocks. Instead, they use a chipping and crushing action with a heavy WOB to penetrate. These bits work more like a chisel on granite than like a shovel in dirt to gouge out smaller pieces of rock than soft-formation bits do.

The cones on medium-hard formation bits have only a slight offset to reduce the twisting action that a lot of offset produces. The cutters are shorter, stronger, and closer together so that each cutter takes a lower load. Typically, three or four rows of cutters exist on each cone. The inserts in bits for medium-hard formations are cone-shaped. The ventilation grooves between the rows of inserts are shallower for harder formations.

Steel-tooth bits for abrasive and hard formations are very short and strong. They often have tungsten carbide hardfacing on the teeth and the gauge surfaces to fortify them against wear. Inserts for the hardest formations are hemispherical on the top and do not project far beyond the cone.

Natural diamond and TSP bits drill formations that are moderately to extremely hard, abrasive, and elastic. Generally, smaller ridge-set diamonds set in a closely spaced pattern drill faster in harder formations. They have greater impact resistance because the load on each diamond is lower. Larger but fewer surface-set diamonds work better in softer (but still moderately hard) formations, where the risk of breaking caused by impact is not as great. They sit out farther from the body of the bit, so they can plow more rock at each rotation, and cleaning and removing cuttings is easier.

Because of the small exposure of the diamonds in either case—remember, a one-carat diamond is only about ¼ inch (6 millimetres) in diameter, so, at most, only $1/12$ inch (2 millimetres) of the diamond is exposed—the bit cuts much less rock during each rotation than a roller cone or PDC bit. The rate of penetration for natural diamond bits is therefore slow. The grade of diamond used also affects the ROP. Round stones are more resistant to impact but cut more slowly. Sharper, high-grade stones cut faster but break more easily.

Formations that are soft to medium-hard with low-to-medium abrasiveness are ideal for PDC bits. PDC bits are very versatile. Their ROP is about three times that of roller cone bits. The stud type penetrates faster than the cylinder type for the same reason that larger natural diamonds drill faster than small ones: more of the diamond layer contacts the formation. PDC cutters are much larger than natural diamonds (½ inch, or 13 millimetres, and larger), so their ROP is much faster.

Hybrid bits are effective in heterogeneous formations with hard stringers. The natural diamond, TSP, or tungsten carbide backup cutters or gauge protection withstand abrasion well and protect the PDC cutters in the harder sections. In the softer sections, the PDC cutters do the work at their typical high ROP.

Diamond Bits

To summarize—

Roller cone bit selection—

- For soft formations, long, widely spaced cutters
- For medium-soft formations, slightly shorter cutters with less offset than soft-formation bits
- For medium-hard formations, still shorter cutters and less offset than medium-soft formation bits
- For hard formations, very short and strong cutters with very little offset

Diamond bit selection—

- For moderately-to-extremely hard, abrasive, and elastic formations, natural diamond and TSP bits
- For soft-to-medium-hard formations with low-to-medium abrasiveness, PDC bits
- For heterogeneous formations with hard stringers, hybrid bits

Weight on Bit, Rotary Speed, and Penetration Rate

Getting the best performance from a bit depends on properly adjusting the weight on the bit and the rotary speed. Generally, the higher the rotary speed, the lower the weight on the bit, and vice versa. Tests in the field and in laboratories have shown that the optimum combination of weight and rotary speed varies for soft, medium, and hard formations. A 30 to 40 percent increase in weight on a steel-tooth bit sometimes doubles the rate of penetration.

Usually, however, an operator or contractor does not just want to improve the drilling rate for a short time. Rather, good overall performance—both the rate of penetration and the total footage per bit—is necessary to achieve the minimum cost per foot of hole. Moderate weights and high rotary speeds are best in quickly drilled, soft, nonabrasive formations. Heavier weights and lower rotary speeds, however, are better for slowly drilled, abrasive, harder formations.

Roller Cone Bits

The method of determining the best weight and rotary speed is the same for steel-tooth and insert bits. The WOB in softer formations usually must be relatively light to avoid balling. But the driller can compensate for the reduced weight, and keep a good penetration rate, by using a faster rotary speed and high-velocity drilling mud circulation for good cleaning.

Rotary speeds of 200 to 250 rpm are not uncommon for steel-tooth bits. The abrasiveness of a formation, however, limits the rotary speed much beyond 250 rpm. Very fast speeds coupled with abrasive particles increase the wear on steel teeth and bearings. Also, insert bits require slower speeds because, at very fast speeds, the inserts can break because of impact shocks. Insert bits will run up to 180 rpm.

Roller cone bits in harder formations require a heavy WOB to crush the rock. Remember that the cutters on rock bits are shorter and closer together and do not shovel the formation, but rather penetrate it like a pointed chisel. In theory, the more weight applied to these chisels (up to their breaking point), the better they work. Of course, the heavier weight stresses the cutters and bearings and wears them out sooner. Another reason to increase the weight on bit when using steel-tooth bits is to maintain an economical drilling rate as the teeth become duller. A heavier weight allows the bit to work longer before the driller must replace it.

Running times vary with the formation and type of bit as well. A steel-tooth bit may last 30 hours in a soft formation. A tungsten carbide insert bit can last as long as 300 hours. Sometimes, therefore, only one or two tungsten carbide bits can drill a well to final depth.

Diamond bits require a lighter WOB than roller cone bits, but the driller still adjusts the WOB and rotary speed to obtain the lowest cost per foot (metre) of hole drilled. The specifications that the manufacturer supplies with a bit will recommend a range for the rotary speed.

PDC bits can drill very fast—up to 200 feet (60 metres) per hour—in soft, nonabrasive formations because the cutters shear deeply into the rock on each rotation. Matrix-body PDC bits resist erosion well and typically run for 300 hours or more. Occasionally, in soft formations, one bit may drill several thousand feet (metres) in each of several wells. PDC rotary speed is generally between 80 and 160 rpm. A driller may rotate a bit made up on positive displacement tool as fast as 600 rpm, or one made up on a turbine as fast as 1,000 to 1,200 rpm.

Natural diamond bits have the slowest penetration rate, 5 to 15 feet (1.5 to 4.5 metres) per hour, because they are grinding very hard formations and drill only a small amount of rock on each pass. They use the lightest weight on bit and relatively high rotary speeds, usually 120 to 140 rpm. Because they drill slowly and because they are often used in the most abrasive formations, they typically last for only 1,000 to 2,000 feet (300 to 600 metres). Keep in mind, however, that even if an expensive diamond bit drills only 50 feet (15 metres) in a given formation, it still performs better than a roller cone bit that would have drilled only 10 feet (3 metres) in the same formation before it wore out.

TSP bits can run up to 30 feet (9 metres) per hour, which is faster than natural diamond bits but does not approach the rate for PDC bits. The WOB and rotary speed for TSP bits fall between those used for PDC and natural diamond bits.

Diamond Bits

To summarize—

- Bit performance depends on properly adjusting WOB and rotary speed.
- When drilling soft formations with both steel-tooth and insert bits, maintain good hydraulics and use a relatively light WOB and a relatively high rpm
- When drilling hard formations with roller cone bits, use a relatively high WOB
- Diamond bits require lighter WOB than roller cone bits
- PDC bits can drill very fast, while natural diamond bits tend to drill more slowly
- TSP bits generally drill faster than natural diamond bits but more slowly than PDC bits

Bit Classification

The IADC has developed a standard system to classify bits. Every bit manufactured anywhere in the world has a classification code based on this system. By reading the code, the driller can evaluate bits from different manufacturers and select the bit needed for a particular job.

Table 1 is a chart that the manufacturer fills in to classify a roller cone bit. The code for roller cone bits has four characters:

1. *Series* of the cutting structure. The first character is a number (called the series) from 1 to 8 that describes the type of bit and the formation it can drill. Series 1 to 3 are for steel-tooth bits and series 4 to 8 refer to tungsten carbide insert bits. Within the two groups, the lowest number stands for bits that drill the softest formations. So a series 1 bit is a steel-tooth bit for a soft formation, and a series 4 bit is an insert bit for a soft formation. Bits for soft formations will have longer and more widely spaced cutters, thinner cones with a greater offset, and no hardfacing.

2. *Type* of cutting structure. The system divides each series into four types to further define the degree of formation hardness that the bit can drill. Type 1 refers to the softest formation within the series. Type 4 is for the hardest.

3. *Bearing type and gauge surface.* This character is a number from 1 to 7 that describes the bearings and what type of gauge protection the bit has. Codes 1 to 5 refer to roller bearings and 6 and 7 refer to journal bearings. The chart divides bearing types into sealed, nonsealed, and air-cooled (also nonsealed) bearings. Codes 3, 5, and 7 apply to bits with extra gauge protection. (Incidentally, the IADC has reserved codes 8 and 9 for future designs, even though they are not shown in table 1.)

4. *Features available.* The last character in the code, which is optional, is one of 16 letters that indicate special features a bit may have. A particular bit may have more than one of these special features, but the manufacturer uses the code for the most important one.

Roller Cone Bits

Table 1
IADC Classification for Steel Tooth and Insert Bits

Manufacturer _____   Date _____

	SERIES	TYPES	FORMATIONS	Standard Roller Bearing (1)	Roller Bearing Air Cooled (2)	Roller Bearing Gauge Protected (3)	Sealed Roller Bearing (4)	Sealed Roller Brg. Gauge Protected (5)	Sealed Friction Bearing (6)	Sealed Friction Brg. Gauge Protected (7)
STEEL-TOOTH BITS	1	1	Soft Formations with Low Compressive Strength and High Drillability							
		2								
		3								
		4								
	2	1	Medium to Medium Hard Formations with High Compressive Strength							
		2								
		3								
		4								
	3	1	Hard Semiabrasive and Abrasive Formations							
		2								
		3								
		4								
INSERT BITS	4	1	Soft Formations with Low Compressive Strength and High Drillability							
		2								
		3								
		4								
	5	1	Soft to Medium Formations with Low Compressive Strength							
		2								
		3								
		4								
	6	1	Medium-Hard Formations with High Compressive Strength							
		2								
		3								
		4								
	7	1	Hard Semiabrasive and Abrasive Formations							
		2								
		3								
		4								
	8	1	Extremely Hard and Abrasive Formations							
		2								
		3								
		4								

FEATURES AVAILABLE

- A - Air Application
- B - Special Bearing Seal
- C - Center Jet
- D - Deviation Control
- E - Extended, Full-Length Jets
- G - Additional Gauge and Body Protection
- H - Horizontal or Steering Application
- J - Jet Deflection
- L - Lug Pads
- M - Motor Application
- S - Standard Steel Tooth Model
- T - Two-Cone Bit
- W - Enhanced Cutting Structure
- X - Predominantly Chisel-Tooth Insert
- Y - Conical Tooth Insert
- Z - Other Shape Insert

1) Several features may be available on any particular bit. The fourth character should describe the predominant feature.
2) All bit types are classified by relative hardness only and will drill effectively in other formations.
3) Please check with the specific bit supplier for additional information.

Diamond Bits

The IADC adopted a system for classifying natural diamond, PDC, and TSP bits in 1990. The codes for these bits have four characters.

1. *Body material.* For PDC bits (table 2) the first character in the code is either M or S, for matrix or steel body.

2. *Cutter density.* The second character is a number from 1 to 4 or from 5 to 8 (tables 2 and 3) that represents the total number of cutters on the bit. A PDC bit (table 2) uses the codes 1 to 4, with 1 meaning the fewest cutters and 4 the most. A natural diamond or TSP bit (table 3) uses codes 6 to 8, which refer to the size of the stones and, by inference, how many there are. Code 6 represents sizes of more than three stones per carat, 7 represents three to seven stones per carat, and 8 represents more than seven stones per carat. In general, bits designated 1 drill the softest formations and 8 the hardest.

3. *Cutter size or type.* The third character is a number from 1 to 3, or, in one case, 4, and depends on the type of bit. For a PDC bit (table 2), the number refers to the size of the cutter, where 1 is the largest and 3, the smallest. For natural diamond and TSP bits (table 3), the codes indicate the type of bit. Code 1 indicates natural diamonds, 2 indicates TSPs, 3 represents hybrid, or combination, bits, and 4 applies to impregnated diamond bits. The IADC *Drilling Manual* defines hybrid bits as those that use natural diamonds and TSPs.

4. *Body style.* The final character is a number from 1 to 4 (across the top of tables 2 and 3) that gives a general idea of the profile. Bits with the shortest flank or side dimension have the designation 1. Codes 2, 3, and 4 represent increasingly longer profiles.

The IADC classification system gives only approximate information about the bit and does not describe the hydraulics at all. It is, however, a simple and functional starting point for comparing bits from different manufacturers.

Table 2
IADC Classification Chart for PDC Bits

(M) = MATRIX BODY (S) = STEEL BODY

CUTTERS DENSITY	SIZE	1 FISHTAIL EC	DBS	HYC	STC	SEC	2 SHORT EC	DBS	HYC	STC	SEC	3 MEDIUM EC	DBS	HYC	STC	SEC	4 LONG EC	DBS	HYC	STC	SEC
1	1 >24	R522(M) R573(M) R523(M)				B943(M)					B17-4(M)										
1	2 14-24		PD12(S)	DS40(S) DS33(S)	S95(S)	B933(M)															
1	3 <14	R423(M) AR423(M)	PD10(S) PD11(S)		S98(S) S93(M)	B923(M)										B254(M)	R516(M)				
2	1 >24	R525(M)							DS30(S)					DS34(S)							
2	2 14-24	R526(M)	TD19L(M)			B925(M)				S25(S)											
2	3 <14	R426(M) Z426(M)	TD2A1(M)	DS39(M)	S93(M)	B935(M)	R482(M)	PD1(S)	DS46(S)	S10(S)	HZ232(M) B2S(M)		LX201(M) LX101(M)	DS26(S) DS31(S)	S45(S)						
3	1 >24											R535(M)									
3	2 14-24		TD19M(M)			B927(M)						R535S(M)	PD4(S)								
3	3 <14		TD5A1(M)				AR435(M)	TD268(M) TD260(M)	DS23(S) DS49(M)		MX42(M)	R435(M)	PD2(S)		S85(S) S43(S)	B272(M)	Z528(M)				
4	1 >24							PD5(S)					PD4HS(S)								
4	2 14-24		TD19H(M)																		
4	3 <14							TD290(M)			HZ352(M) B352(M)	R437(M) Z437(M)	LX401(M) LX301(M)	D247(M)	S35(M)	S292(M)	R419(M) R428(M) Z428(M)	LX271(M) TD115(M) LX291(M)	DS18(M) DS19(M) DS29(M)		B102(M) B362(M)

BODY STYLE

JC001

Table 3
IADC Classification Chart for TSP and Natural Diamonds

CUTTER SIZE	#	ELEMENT	1 FLAT EC	DBS	HYC	STC	SEC	2 SHORT EC	DBS	HYC	STC	SEC	3 MEDIUM EC	DBS	HYC	STC	SEC	4 LONG EC	DBS	HYC	STC	SEC
6 <3 SPC	1	NAT											D262 D311	TB16	901 932		N37	D18			N42	
	2	TSP						S725					S225	TT16	211 241							
	3	COMB												TBT16	211ND 241ND							
7 3-7 SPC	1	NAT	D411	TB26	828		N4S	D41	TB521				D262 D331 D311	TB601	901 730 753 744		N39 N50	T51 T54	TB593 TB703	901DT		
	2	TSP	SST		828TSP				TT521	263		P443	S248 S226	TT601	243 223		P341 P343		TT593			
	3	COMB							TBT521	263ND				TBT601	243ND 223ND				TBT593 TBT703			
8 >7 SPC	1	NAT						D24		525 585		N60										
	2	TSP																				
	3	COMB																				
JC002	4	IMP	S279						TB5211													

BODY STYLE

To summarize—

- Roller cone bits are classified according to their cutting structure, bearing design, gauge protection, and other features peculiar to a particular design.
- Diamond bits are classified according to their body material, cutter density, cutter size and type, and profile.

Dull Bit Grading

It is very important to grade dull bits properly. Grading a dull bit means estimating how much and where it has worn. Proper dull bit grading helps the operator and the contractor correct poor drilling practices, select the best type of bit for specific conditions, and make decisions that affect the cost of future drilling. It is a form of ongoing field testing that benefits all drilling contractors and operators.

Roller cone bits and fixed-head (diamond) bits are both graded using an IADC dull bit classification system with eight categories (table 4). Since fixed-head bits have no bearings, the column for bearing wear (B) always has an X in it when grading diamond bits. The list of codes for cutter wear includes some that apply only to diamond bits and some that apply only to roller cone bits.

The first four columns on the dull bit grading chart refer to the condition of the cutters. Column 1 refers to the inner two-thirds of the cutters, the ones that do not touch the side of the hole. Column 2 refers to the outer one-third of the cutters (the gauge cutters). Each of these columns can have a grade from 0 to 8 that indicates the number of cutters left on the bit and how much they are worn. Code 0 represents no wear, and 8 means that no usable cutters are left. A grade of 4 means that 50 percent of the cutters remain usable. Column 3 ("dull char.") contains specific information about how the cutters are worn. There are many two-letter codes for various types of wear, as the table shows. Column 4 ("location") shows the exact location of the worn cutters.

Table 4
IADC Dull Grading Chart

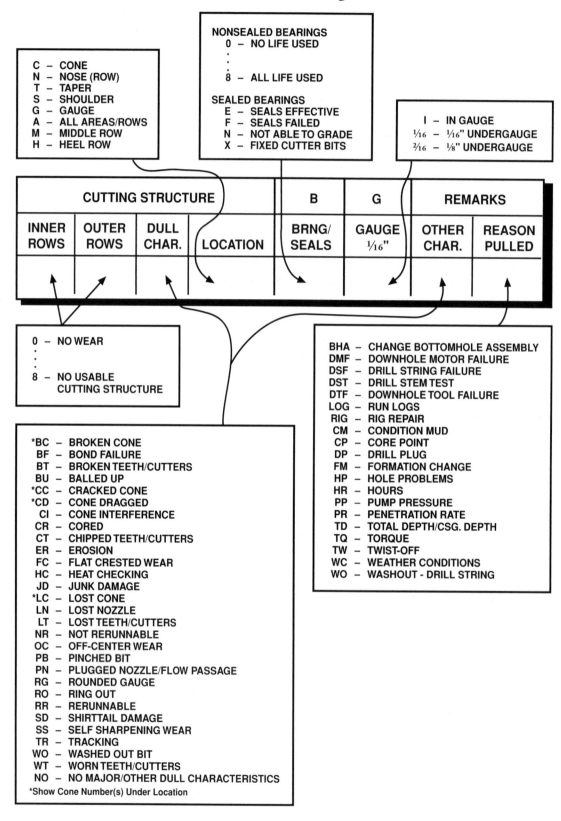

C – CONE
N – NOSE (ROW)
T – TAPER
S – SHOULDER
G – GAUGE
A – ALL AREAS/ROWS
M – MIDDLE ROW
H – HEEL ROW

NONSEALED BEARINGS
0 – NO LIFE USED
.
.
8 – ALL LIFE USED

SEALED BEARINGS
E – SEALS EFFECTIVE
F – SEALS FAILED
N – NOT ABLE TO GRADE
X – FIXED CUTTER BITS

I – IN GAUGE
1/16 – 1/16" UNDERGAUGE
2/16 – 1/8" UNDERGAUGE

CUTTING STRUCTURE				B	G	REMARKS	
INNER ROWS	OUTER ROWS	DULL CHAR.	LOCATION	BRNG/ SEALS	GAUGE 1/16"	OTHER CHAR.	REASON PULLED

0 – NO WEAR
.
.
8 – NO USABLE
    CUTTING STRUCTURE

*BC – BROKEN CONE
 BF – BOND FAILURE
 BT – BROKEN TEETH/CUTTERS
 BU – BALLED UP
*CC – CRACKED CONE
*CD – CONE DRAGGED
 CI – CONE INTERFERENCE
 CR – CORED
 CT – CHIPPED TEETH/CUTTERS
 ER – EROSION
 FC – FLAT CRESTED WEAR
 HC – HEAT CHECKING
 JD – JUNK DAMAGE
*LC – LOST CONE
 LN – LOST NOZZLE
 LT – LOST TEETH/CUTTERS
 NR – NOT RERUNNABLE
 OC – OFF-CENTER WEAR
 PB – PINCHED BIT
 PN – PLUGGED NOZZLE/FLOW PASSAGE
 RG – ROUNDED GAUGE
 RO – RING OUT
 RR – RERUNNABLE
 SD – SHIRTTAIL DAMAGE
 SS – SELF SHARPENING WEAR
 TR – TRACKING
 WO – WASHED OUT BIT
 WT – WORN TEETH/CUTTERS
 NO – NO MAJOR/OTHER DULL CHARACTERISTICS
*Show Cone Number(s) Under Location

BHA – CHANGE BOTTOMHOLE ASSEMBLY
DMF – DOWNHOLE MOTOR FAILURE
DSF – DRILL STRING FAILURE
DST – DRILL STEM TEST
DTF – DOWNHOLE TOOL FAILURE
LOG – RUN LOGS
RIG – RIG REPAIR
CM – CONDITION MUD
CP – CORE POINT
DP – DRILL PLUG
FM – FORMATION CHANGE
HP – HOLE PROBLEMS
HR – HOURS
PP – PUMP PRESSURE
PR – PENETRATION RATE
TD – TOTAL DEPTH/CSG. DEPTH
TQ – TORQUE
TW – TWIST-OFF
WC – WEATHER CONDITIONS
WO – WASHOUT - DRILL STRING

Column 5 (B) refers to the bearing wear. The grading of a used bearing is the most difficult because it must be an estimate. Unlike cutters, where the driller can measure the remaining height and compare it to the known height when new, there is no system for measuring the wear to bearings. Only an experienced driller who knows how many hours the bit drilled and the operating conditions can make such an estimate. Nonsealed bearings get a code from 0 to 8. Code 0 indicates a new bearing, and 8 indicates that the bearing is useless (locked or missing). A code of 6 means that ⁶⁄₈ of the estimated life of the bearing has been used. Sealed bearings get a letter E, F, or N. E means the seal is effective, F means the seal has failed, and N means that the driller was not able to grade the seal or the bearing condition.

Column 6 (G) shows whether the bit can still drill the size of hole (the gauge) it should. The letter I means that the bit is still in gauge. If the bit is undergauge, record the amount in sixteenths of an inch (or in millimetres). To determine this measurement for a roller cone bit with three cones, use a ring gauge and the two-thirds rule. Place the ring gauge over the bit and hold it so that two of the cones touch the ring at their outermost points. If the bit is undergauge, the ring gauge will not touch the third cone—a gap will exist between the inside of the ring gauge and the outside surface of the cone. Measure the distance between the third cone and the ring gauge and multiply by ⅔ (0.666) (fig. 85). Round that number to the nearest ¹⁄₁₆ of an inch (millimetre) to get the code for wear in this column.

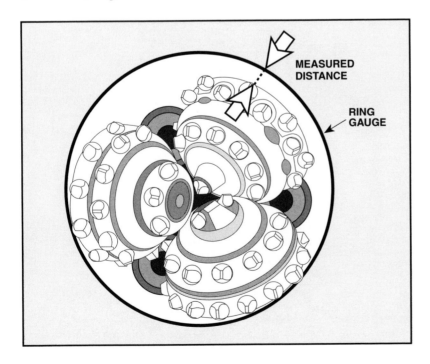

Figure 85. Ring gauge held so that two cones touch the ring

Column 7 in table 4 contains two-letter codes for any secondary wear to either the cutters or the bit as a whole. The list in column 3 contains these codes.

Finally, column 8 is a two- or three-letter code for the reason the driller pulled the bit from the hole.

To summarize—

- Roller cone bits and diamond bits use the same dull bit grading system. They are graded on the basis of cutter wear, bearing wear (not for diamond bits), and gauge wear.

Costs

Drilling contractors often use a break-even analysis to determine the cost of drilling per foot (metre) of hole. They take into account the following expenses:

- Cost to operate the rig, per hour
- Cost of the bit
- Trip time, in hours
- Drilling time, in hours

The choice of bit affects all of these expenses. A PDC bit may cost 15 times as much as the same-sized steel-tooth bit and four times as much as the same-sized tungsten carbide insert bit. The initial cost of the bit is only one factor, however. Using a PDC bit affects other aspects of drilling, such as how long it takes to drill to a certain depth and how many time-consuming trips a hole will need (PDC bits last longer). The longer the drilling time, the more the drilling contractor has to pay for operating the rig—such things as fuel, repairing or replacing equipment, and wages. Since drilling companies intend to make a profit, the widespread use of PDC bits shows that they must be economical in many situations (fig. 86).

The price of hybrid bits varies depending on the features, but they are at least as expensive as ordinary diamond bits. Hybrid bits may be, however, the most economical bits for hard formations. A recent study determined that the more hostile the formation, the greater the savings in cost per foot of using a hybrid bit.

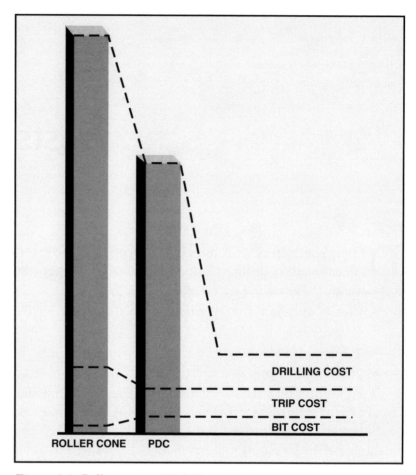

Figure 86. Roller cone vs PDC bits

To summarize—

To determine the cost of drilling a foot (metre) of hole, contractors include—

- Cost to operate the rig, per hour
- Cost of the bit
- Trip time, in hours
- Drilling time, in hours

Field Operating Procedures

▼
▼
▼

lthough a roller cone bit is about as foolproof as any piece of equipment on the drilling rig, if the driller abuses it, its life is cut short. Certain general procedures can help prolong its life. Pointers for making up the bit include the following:

Drilling With a Roller Cone Bit

- If the box containing the bit is open, check the threads on the bit shank and clean them if needed.
- Lubricate the threads with a good-quality, clean lubricant intended for tool joint threads.
- Use the correct breaker plate, or bit breaker, for the size and type of bit.
- Cover the hole and place the breaker in the locked rotary table. Screw the bit into the box threads on the collar sub. Place the bit in the breaker, and lower the collar over the shank. Rotate the collar by hand until it fits down over the bit.
- Place the makeup tongs on the collar just above the bit. Make up to the proper torque. Never use a sledgehammer on a bit or on the tongs to tighten the bit.

When tripping in, work the bit slowly past any ledges. If the hole is slightly undergauge, ream it out carefully. As you approach the bottom, run the bit in with full circulation to clean out cuttings that have settled on the bottom. Rushing this step can cause the bit to ball up and the cones to start dragging.

Drilling with a Diamond Bit

Because of their unique strengths and limitations, diamond bits, with either synthetic or natural diamond cutters, require different drilling techniques from roller cone bits. With diamond bits, it is particularly important to adjust the weight on bit, the rotary speed, and the hydraulics for the most economical drilling and the most efficient wear of the bit.

First, prepare the hole for the new bit. Look at the wear on the previous bit used in the hole; look especially for missing cutters and damage from junk. These indicate that it may be necessary to run a junk basket or other junk fishing device to clean the hole before running the new bit. It is very important to make sure the hole is clean before making up the new bit because junk can break the diamonds.

Next, bring the bit to the rig floor still inside its packing box to avoid damaging the cutters. Handle the bit carefully. Do not set it directly on a steel floor, but on a wooden pad or rubber mat. Inspect the cutters of the new bit for damage. Check that the nozzles are of the correct size, then check the gauge. Record the bit's serial number. Finally, look for any foreign matter inside the bit.

When tripping in, move ahead slowly past any ledges, doglegs, or tight spots until the bit just tags (touches) the bottom. Then raise it off bottom about a foot, start rotating, and circulate with full flow for 5 to 10 minutes at 50 to 60 rpm. To break in the bit, lower it to bottom with full flow at 60 to 80 rpm and use a light weight on the bit. Establish the bit's bottomhole pattern by drilling slowly with a light WOB. Continue drilling this way for 1 to 3 feet (1 metre). Then increase the rotary speed and add weight in 2,000-pound (1,000-decanewton) increments to reach the optimum weight on the bit. Finally, while maintaining optimum WOB, vary the rotary speed to find the best rate of penetration.

While drilling ahead, do not hesitate to adjust the rotary speed, weight on the bit, pump pressure, and so on to maintain the best ROP as the bit wears. Pay particular attention to the formation hardness, and reduce the rotary speed, if necessary, to keep the temperature at the cutters from getting too high.

To pull out of the hole, be as careful as when tripping in. Move slowly past tight spots, and handle the bit carefully. Inspect and grade the dull bit before returning it to its box.

To summarize—

To determine the cost of drilling a foot (metre) of hole, contractors include—

- Cost to operate the rig, per hour
- Cost of the bit
- Trip time, in hours
- Drilling time, in hours

To run a roller cone bit—

- Make it up correctly on the drill collar
- As the bit approaches bottom, use full circulation to clean the bottom of the hole

To run a diamond bit—

- Make sure the hole is full gauge; if not ream it to full gauge
- Make sure the hole is clean
- Run the bit to bottom with full circulation
- To break in the bit, lower it to bottom with full circulation at 60–80 rpm; use light WOB and drill about 1–3 feet (1 metre) of hole
- Increase the rotary speed and add weight in 2,000-pound (1,000-decanewton) increments to reach optimum WOB
- Maintain optimum WOB and vary the rotary speed to find the best ROP
- Be careful when tripping out the bit; move slowly past tight spots, and handle the bit carefully

Conclusion

▼
▼
▼

The properties of the metals and diamonds used in bits, the hydraulic considerations, the properties of the formations being drilled, and the weight and rotary speed applied to the bit all influence each other in a complex way. It is not possible to fully understand one part without understanding all the others. For this reason, it may help you to read this manual again. You will find, for example, that you can make better sense of bit designs now that you know more about formations. The more you understand about these complex interactions, the more valuable you will be to the drilling contractor.

Glossary

abrasion *n*: wearing away by friction.

air bit *n*: a roller cone bit that is specially designed for air or gas drilling. It is very similar to a regular bit, but features screens over the bearings to protect them from clogging with cuttings and thicker hardfacing on the shirttail to protect them from abrasive, high-velocity air or gas drilling fluid.

air drilling *n*: a method of rotary drilling that uses compressed air, instead of water or mud, as the circulating medium; called gas drilling if compressed natural gas instead of air is circulated.

alloy *n*: a substance with metallic properties that comprises two or more elements in solid solution.

annular velocity *n*: the speed at which the drilling fluid is traveling in the annulus of a well.

annulus *n*: the space between the drill string and the wall of the hole or the casing.

antiwhirl bit *n*: a specially designed fixed-head bit with a pad on the bit's gauge surface that unbalances the bit and moves the center of rotation to the pad.

back rake angle *n*: in a PDC bit, the amount of distance, in degrees, between vertical (90°) and the face of the cutter.

ball bearing *n*: a bearing made up of several relatively small steel spheres that rotate freely in a groove, or race, and convert sliding friction into rolling friction. Compare *journal bearing, roller bearing*.

balled-up bit *n*: a bit on which a mass of sticky consolidated material, usually drilled cuttings, has accumulated.

ball race *n*: a track, channel, or groove in which ball bearings turn.

barite *n*: barium sulfate, $BaSO_4$; a mineral frequently used to increase the weight, or density, of drilling mud.

bearing *n*: see *ball bearing*.

bearing pin *n*: a cylindrical projection from each leg of a bit onto which the bit's bearings are mounted.

bed *n*: a specific layer of earth or rock that presents a contrast to other layers of different materials lying above, below, or adjacent to it.

bentonite *n*: a colloidal clay, composed primarily of montmorillonite, that swells when wet. Because of its gel-forming properties, bentonite is a major component of water-base drilling muds.

bent sub *n*: a short cylindrical device installed in the drill stem between the bottommost drill collar and a downhole motor. Its purpose is to deflect the downhole motor off vertical to drill a directional hole.

bit *n*: the cutting or boring element used in drilling oil and gas wells. The bit consists of a cutting element and a circulating element. The cutting element is either steel teeth, tungsten carbide buttons, industrial-grade natural diamonds, synthetic diamonds, or polycrystalline diamonds (PDCs). These teeth, buttons, or diamonds penetrate and gouge or scrape the formation to remove it. The circulating element permits the passage of drilling fluid and utilizes the hydraulic force of the fluid stream to improve drilling rates.

bit breaker *n*: a heavy plate that fits in the rotary table and holds the drill bit while the crew makes it up or breaks it out of the drill stem. See also *bit*.

bit cone *n*: see *roller cone bit*.

bit gauge *n*: a circular ring used to determine whether a bit is of the correct outside diameter.

bit hydraulic horsepower *n*: the measure of hydraulic power expended through the bit nozzles for cleaning the bit cutters and the hole bottom.

bit matrix *n*: on a diamond bit, the material (usually powdered and fused tungsten carbide) into which the diamonds are set.

bit pin *n*: the threaded element at the top of a bit that allows it to be made up in a drill collar or other component of the drill stem.

bit plot *n*: on diamond bits, the pattern in which the diamonds are placed in the face of the bit.

bit profile *n*: in fixed-head bits, the shape of the cross section of the body of the bit.

bit program *n*: a plan for the expected number and types of bits that are to be used in the drilling of a well. The bit program takes into account all the factors that affect bit performance so that reliable cost calculations can be made.

bit record *n*: a report that lists each bit used during a drilling operation, giving the type, the footage it drilled, the formation it penetrated, its condition, and so on.

bit run *n*: the placing of a bit on the bottom of the hole, drilling with it until it drills the prescribed amount of hole, or until it wears out, and pulling it out of the hole.

bit shank *n*: the threaded portion of the top of the bit that is screwed into the drill collar. Also called the pin.

bit sub *n*: a sub inserted between the drill collar and the bit. See *sub*.

bit whirl *n*: the motion a bit makes when it does not rotate around its center but instead drills with a spiral motion. It usually occurs to a bit drilling in a soft or medium soft formation when the driller does not apply enough weight or does not rotate the bit fast enough. A whirling bit drills an overgauge hole (a hole larger than the diameter of the bit) and causes the bit to wear abnormally.

blowout *n*: the uncontrolled flow of fluids, such as gas, oil, and water, from a wellbore.

boot sub *n*: a device made up in the drill stem above the mill to collect bits of junk ground away during a milling operation. During milling, drilling mud under high pressure forces bits of junk up the narrow space between the boot sub and hole wall. When the junk reaches the wider annulus above the boot sub and pressure drops slightly, the junk falls into the boot sub. A boot sub also can be run above the bit during routine drilling to collect small pieces of junk that may damage the bit or interfere with its operation. Also called a junk sub or junk boot.

bottomhole assembly *n*: the portion of the drilling assembly below the drill pipe. It can be very simple—composed of only the bit and drill collars—or it can be very complex and made up of several drilling tools.

box *n*: the female section of a connection. See *tool joint*.

box and pin *n*: see *tool joint*.

box threads *n pl*: threads on the female section, or box, of a tool joint. See *tool joint*.

bradding *n*: a condition in which the weight on a bit tooth has been so great that the tooth has dulled until its softer inner portion caves over the harder case area.

break-in *n*: the drilling of the first few feet of hole with a new bit.

broken cone *n*: a cone on a bit that has become cracked.

button bit *n*: see *tungsten carbide bit*.

C

carbonate rock *n*: a sedimentary rock composed primarily of calcium carbonate (calcite) or calcium magnesium carbonate (dolomite).

Celsius scale *n*: the metric scale of temperature measurement used universally by scientists. On this scale, 0° represents the freezing point of water and 100° its boiling point at a barometric pressure of 760 millimetres. Degrees Celsius are converted to degrees Fahrenheit by using the following equation:

$$°F = \tfrac{9}{5}(°C) + 32.$$

The Celsius scale was formerly called the centigrade scale; now, however, the term "Celsius" is preferred in the International System of Units (SI).

center coring *n*: a condition on a bit in which the inside row of cutters on the cones wears or breaks or the nose of one or more of the cones breaks.

center jet *n*: on roller cone bits, especially those designed for drilling soft, sticky formations, an additional jet nozzle that is added above the cones and in the middle of the bit. In some cases, the center jet, in addition to the three jets on the sides of the bit, can help keep the bit cutters clean.

centimetre (cm) *n*: a unit of length in the metric system equal to one-hundredth of a metre (10^{-2} metre).

chatter *n*: a phenomenon in which the cutters of a bit, while rotating on bottom, do not adequately penetrate the formation; instead, the cutters tend to bounce off the formation. The bounce is sometimes transmitted to the drill stem in the form of a vibration, or chatter.

chip hold-down effect *n*: the holding of formation rock chips in place as a result of high differential pressure in the wellbore (i.e., pressure in the wellbore is greater than pressure in the formation). This effect limits the cutting action of the bit by retarding circulation of bit cuttings out of the hole.

circle plot *n*: in diamond bits, a pattern of setting diamonds in which the manufacturer spreads them in concentric circles on the cutting surface (the nose) of the bit. Compare *grid plot*.

circulate *v*: to pass from one point throughout a system and back to the starting point. For example, drilling fluid is circulated out of the suction pit, down the drill pipe and drill collars, out the bit, up the annulus, and back to the pits while drilling proceeds.

clastic rock *n*: a sedimentary rock composed of fragments of preexisting rocks. The principal distinction among clastics is grain size. Conglomerates, sandstones, and shales are clastic rocks.

compact *n*: any device made of metallic, or other, powders pressed together very hard to form a solid mass. Polycrystalline diamond compact (PDC) bits have cutters that consist of several compacts fitted into the body of the bit.

compressive strength *n*: a measure of a material's ability to withstand heavy weights, or loads, pressing on them from above or below before failure occurs.

condition *v*: to treat drilling mud with additives to give it certain properties. To condition and circulate mud is to ensure that additives are distributed evenly throughout a system by circulating the mud while it is being conditioned.

cone *n*: a conical-shaped metal device into which cutting teeth are formed or mounted on a roller cone bit. See *roller cone bit*.

cone alignment *n*: on a roller cone bit, the way in which the cones of a bit line up with the bit's center axis. If the cones line up with the center axis, they have on-center alignment; if they do not line up with the center axis, they have off-center alignment. Generally, bits with on-center alignment drill hard formations; those with off-center alignment drill soft formations.

cone bit *n*: a roller bit in which the cutters are conical. See *bit*.

cone erosion *n*: a phenomenon in which a bit cone is worn away (eroded) by abrasive materials in the drilling fluid or from other sources.

cone interference *n*: a problem on a bit in which one cone interferes with the action of the other cones. Interference can occur when the bit is jammed into an undergauge (undersize) borehole; the legs of the bit are forced inward, which causes the cones to interfere with each other.

cone offset *n*: the amount by which lines drawn through the center of each cone of a bit fail to meet in the center of the bit. For example, in a roller cone bit with three cones, three lines can be drawn through the center of each cone and extended to the center of the bit. If these cone centerlines do not meet in the bit's center, the cones are said to be offset. In general, bits designed for drilling soft formations have more offset than cones for hard formations, because offset affects the angle at which the bit teeth contact the formation. Since soft formations require a gouging and scraping action by bit teeth, high offset achieves the necessary action.

cone shake *n*: shaking or vibrating of the cones of a bit that occurs when the bit bearings are worn.

cone shell *n*: that part of the cone of a roller cone bit out of which the teeth are milled or into which tungsten carbide inserts are placed and inside of which the bearings are housed.

cone skidding *n*: locking of a cone on a roller cone bit so that it will not turn when the bit is rotating. Cone skidding results in the flattening of the surface of the cone in contact with the bottom of the hole. Also called cone dragging.

conical angle *n*: the angle of the cone of a bit. This angle may be steep, in which case the cone has a sharp taper, or it may be shallow, in which case the cone has a flatter taper.

core bit *n*: a bit that does not drill out the center portion of the hole, but allows this center portion (the core) to pass through the round opening in the center of the bit and into the core barrel.

correlate *v*: to relate subsurface information obtained from one well to that of others so that the formations may be charted and their depths and thicknesses noted. Correlations are made by comparing electrical well logs, radioactivity logs, and cores from different wells.

cracked cone *n*: a bit cone that has been cracked open for any reason.

cross-pad flow *n*: see *feeder collector*.

crow's foot *n*: on a diamond bit, the built-in channels in the bit's nose that helps conduct drilling fluid exiting the center of the bit to additional channels that conduct the fluid over the entire nose.

cutters *n pl*: cutting teeth or other devices on the cones of a roller cone bit or in the body of a diamond bit.

cuttings *n pl*: the small pieces of rock generated by a bit's cutters when drilling.

D

deflection *n*: a change in the angle of a wellbore. In directional drilling, it is measured in degrees from the vertical.

deflection tool *n*: a device made up in the drill string that causes the bit to drill at an angle to the existing hole. It is often called a kickoff tool, because it is used at the kickoff point to start building angle.

delamination *n*: in a polycrystalline diamond (PDC) bit, the failure of the bond between the tungsten carbide backing and the layer of polycrystalline diamonds. Delamination causes the diamond layer to fall off leaving only tungsten carbide to cut the rock.

deviation *n*: departure of the wellbore from the vertical, measured by the horizontal distance from the centerline of the rotary table to the target. The amount of deviation is a function of the drift angle and hole depth. The term is sometimes used to indicate the angle from which a bit has deviated from the vertical during drilling. See *drift angle*.

diamond bit *n*: a drill bit that has small industrial diamonds embedded in its cutting surface. Cutting is performed by the rotation of the very hard diamonds over the rock surface.

directional drilling *n*: intentional deviation of a wellbore from the vertical. Although wellbores are normally drilled vertically, it is sometimes necessary or advantageous to drill at an angle from the vertical. Controlled directional drilling makes it possible to reach subsurface areas laterally remote from the point where the bit enters the earth. It often involves the use of deflection tools.

downhole motor *n*: a drilling tool made up in the drill string directly above the bit. It causes the bit to turn while the drill string remains fixed. It is used most often as a deflection tool in directional drilling, where it is made up between the bit and a bent sub (or, sometimes, the housing of the motor itself is bent). Two principal types of downhole motor are the positive-displacement motor and the downhole turbine motor. Also called mud motor.

drag *n*: friction between a moving device (such as a bit) and another moving or nonmoving part (such as the formation).

drag bit *n*: any of a variety of drilling bits that have no moving parts. As they are rotated on bottom, elements of the bit make hole by being pressed into the formation and being dragged across it. Also called a fishtail bit.

drift angle *n*: the angle at which a wellbore deviates from vertical.

drill bit *n*: see *bit*.

driller's log *n*: a record that describes each formation encountered as it is drilled and lists the drilling time relative to depth, usually in 5- to 10-foot (1.5- to 3-metre) intervals.

drilling fluid *n*: circulating fluid, one function of which is to lift cuttings out of the wellbore and to the surface. Other functions are to cool the bit and to counteract downhole formation pressure. Although a mixture of barite, clay, water, and chemical additives is the most common drilling fluid, wells can also be drilled by using air, gas, water, or oil-base mud as the drilling mud. See also *mud*.

drilling rate *n*: the speed with which the bit drills the formation; usually called the rate of penetration (ROP).

drill stem *n*: all members in the assembly used for rotary drilling from the swivel to the bit, including the kelly, drill pipe and tool joints, drill collars, stabilizers, and various specialty items. Compare *drill string*.

drill string *n*: the column, or string, of drill pipe with attached tool joints that transmits fluid and rotational power from the kelly to the drill collars and bit. Often, especially in the oil patch, the term is loosely applied to both drill pipe and drill collars. Compare *drill stem*.

Dyna-Drill *n*: trade name for a downhole motor driven by drilling fluid that imparts rotary motion to a drilling bit connected to the tool, thus eliminating the need to turn the entire drill stem to make hole. Used in straight and directional drilling.

eccentric bit *n*: a bit that does not have a uniformly round cross section; instead, the bit has a protuberance that projects from one side. An eccentric bit drills a hole slightly overgauge to compensate for certain formations, such as shale or salt, that deform and enlarge after being drilled. Eccentric bits can also ream out undergauge holes.

E

elasticity *n*: the capability of an object that is put under stress (e.g., is stretched) to recover its size and shape when the stress is released. Compare *plasticity*.

elasticity modulus *n*: see *modulus of elasticity*.

elastic modulus *n*: see *modulus of elasticity*.

elastomer *n*: any of various elastic substances that resemble rubber.

extended nozzle *n*: a special bit nozzle, often used on large bits, that lengthens the nozzle and therefore places the jet of drilling fluid exiting the nozzle close to the bottom of the hole. With large bits, where regular nozzles can be relatively distant from the bottom of the hole, the cleaning power of the jet of drilling fluid may be lost because the velocity, or speed, of the jet diminishes rapidly after it exits the nozzle. By extending the length of the nozzles, the jets are placed closer to bottom for maximum cleaning.

Fahrenheit scale *n*: a temperature scale devised by Gabriel Fahrenheit, in which 32° represents the freezing point and 212° the boiling point of water at standard sea-level pressure. Fahrenheit degrees may be converted to Celsius degrees by using the following formula:

F

$$°C = \tfrac{5}{9} \, (°F - 32).$$

feeder-collector *n*: in diamond bits, the type of flow channel that directs drilling fluid across the cutting surface of the bit in which small channels called feeders direct the fluid to flow into other channels called collectors. Compare *radial flow*.

ferrous alloy *n*: a metal alloy in which iron is a major component.

fish *n*: an object that is left in the wellbore during drilling or workover operations and that must be recovered before work can proceed. It can be anything from a piece of scrap metal to a part of the drill stem. *v*: to recover from a well any equipment left there during drilling operations, such as a lost bit or drill collar or part of the drill string.

fishtail bit *n*: a drilling bit with cutting edges of hard alloys. Developed about 1900, and first used with the rotary system of drilling, it is still useful in drilling very soft formations. Also called a drag bit.

fixed-head bit *n*: a bit, such as a diamond or a PDC bit, whose cutters do not rotate or move as the driller rotates the bit on bottom. Compare *roller-cone bit*.

flange *n*: on a bearing race, the rims around the edge of the race that help retain the bearings in the race.

flank *n*: the sides of a bit.

fluid *n*: a substance that flows and yields to any force tending to change its shape. Liquids and gases are fluids.

friction *n*: the force that resists movement between two objects in contact.

friction bearing *n*: see *journal bearing*.

friction loss *n*: a reduction in the pressure of a fluid caused by its motion against an enclosed surface (such as a pipe). As the fluid moves through the pipe, friction between the fluid and the pipe wall and within the fluid itself creates a pressure loss. The faster the fluid moves, the greater are the losses.

full-gauge bit *n*: a bit that has maintained its original diameter.

full-gauge hole *n*: a wellbore drilled with a full-gauge bit. Also called a true-to-gauge hole.

G

gall *n*: a worn area on the surface of a bearing.

gas drilling *n*: see *air drilling*.

gauge *n*: the diameter of a bit or the hole drilled by the bit. *v*: to measure size, volume, depth, or other measurable property.

gauge area *n*: the outside edges of a bit; those portions of a bit that contact the wall of the hole or parallel the wall of the hole.

gauge cutters *n pl*: the teeth or tungsten carbide inserts in the outermost row on the cones of a bit, so called because they cut the outside edge of the hole and determine the hole's gauge or size. Also called gauge row.

gauge rounding *n*: a phenomenon in which the outside areas (the gauge areas) of a bit become worn and rounded by an abrasive formation.

grid plot *n*: in diamond drilling bits, a pattern of setting diamonds in which the manufacturer spreads them evenly over the cutting surface (the nose) of the bit. Compare *circle plot*.

H

hardfacing *n*: very hard material, such as tungsten carbide, that is applied to the surface of bit cutters, bit shirttails, and other areas for the purpose of preventing wear.

hardness *n*: the resistance of a substance to scratching or to indentation.

heat checking *n*: a condition that occurs when the cutters of a bit drag on a formation and become very hot because of friction and then are cooled by the drilling fluid; the rapid cooling causes small cracks (heat checks) to develop.

heel teeth *n*: see *gauge cutters*.

hole angle *n*: the angle at which a hole deviates from vertical.

hole cleaning *n*: the lifting and removing of cuttings made by a bit's cutters by the drilling fluid coming out the bit.

hole drift *n*: the amount a wellbore is deflected from vertical.

hole geometry *n*: the shape and size of the wellbore.

hybrid bit *n*: a bit that combines the features of natural diamond, polycrystalline diamond compact, thermally stable polycrystalline, and, sometimes, roller cone bits.

hydraulic horsepower (hhp) *n*: a measure of the power of a fluid under pressure—that is, fluid volume plus velocity.

hydraulics *n*: 1. the branch of science that deals with practical applications of water or other liquid in motion. 2. the planning and operation of a rig hydraulics program, coordinating the power of the circulating fluid at the bit with other aspects of the drilling program so that bottomhole cleaning is maximized.

IADC *abbr*: International Association of Drilling Contractors.

I

inner bearings *n pl*: in a bit, the bearings that lie inside of and near the end, or nose, of the cone of a roller cone bit. Typically, the inner bearings include the bearings in the nose of the cone and the bearings near the middle of the cone. These bearings may be a combination of roller, ball, or journal bearings. Compare *outer bearings*.

insert *n*: a cylindrical object, rounded, blunt, or chisel-shaped on one end and usually made of tungsten carbide, that is inserted in the cones of a bit, the cutters of a reamer, or the blades of a stabilizer to form the cutting element of the bit or the reamer or the wear surface of the stabilizer. Also called a compact.

insert bit *n*: see *tungsten carbide bit*.

interfit *n*: the distance that the ends of one bit cone extend into the grooves of an adjacent one in a roller cone bit. Also called intermesh.

intermesh *n*: see *interfit*.

International Association of Drilling Contractors (IADC) *n*: an organization of drilling contractors that sponsors or conducts research on education, accident prevention, drilling technology, and other matters of interest to drilling contractors and their employees. Its official publication is *The Drilling Contractor*. Address: Box 4287; Houston, Texas 77210; (713) 578-7171; fax (713) 578-0589.

International System of Units (SI) *n*: a system of units of measurement based on the metric system, adopted and described by the Eleventh General Conference on Weights and Measures. It provides an international standard of measurement to be followed when certain customary units, both US and metric, are eventually phased out of international trade operations. The symbol SI (Le Système International d'Unités) designates the system, which involves seven base units: (1) metre for length, (2) kilogram for mass, (3) second for time, (4) Celsius for temperature, (5) ampere for electric current, (6) candela for luminous intensity, and (7) mole for amount of substance. From these units, others are derived without introducing numerical factors.

jet bit *n*: a drilling bit having replaceable nozzles through which the drilling fluid is directed in a high-velocity stream to the bottom of the hole to improve the efficiency of the bit. See *bit*.

JKL

jet deflection bit *n*: a special jet bit that has a very large nozzle used to deflect a hole from the vertical. The large nozzle erodes one side of the hole so that the hole is deflected off vertical. A jet deflection bit is especially effective in soft formations.

journal *n*: the part of a rotating shaft that turns in a bearing.

journal angle *n*: the angle formed by lines perpendicular to the axis of the journal and the axis of the bit. Also called pin angle.

journal bearing *n*: a machine part in which a rotating shaft (a journal) revolves or slides. Also called a plain bearing. Compare *ball bearing, roller bearing*.

junk *n*: metal debris lost in a hole. Junk may be a lost bit, pieces of a bit, milled pieces of pipe, wrenches, or any relatively small object that impedes drilling or completion and must be fished out of the hole.

junk slot *n*: a groove in the side of a diamond or PDC bit that creates an area of low velocity where relatively small pieces of junk will be lifted and then sent up the hole in the drilling fluid.

kick *n*: the entry of formation fluids into the wellbore or into the drill stem.

long substrate (LS) bond failure *n*: the failure of the adhesion (bond) between the tungsten carbide stud and the tungsten carbide of a polycrystalline diamond compact (PDC) bit cutter. As a result, the PDC layer and the tungsten carbide layer below the PDC layer fall off, leaving only the tungsten carbide stud on the bit.

lost circulation *n*: a phenomenon in which drilling mud circulated from the surface enters fractures, fissures, or cavernous openings in the formation adjacent to the wellbore. In some cases, only part of the mud is lost into the formation and some or most of it returns up the annulus to the surface; in the worst case, all of the mud flows into the formation and none returns to the surface.

M

matrix *n*: 1. in rock, the fine-grained material between larger grains in which the larger grains are embedded. A rock matrix may be composed of fine sediments, crystals, clay, or other substances. 2. the material in which the diamonds on a diamond bit are set.

metric system *n*: a decimal system of weights and measures based on the metre as the unit of length, the gram as the unit of weight, the cubic metre as the unit of volume, the litre as the unit of capacity, and the square metre as the unit of area.

mill *n*: a downhole tool with rough, sharp, extremely hard cutting surfaces for removing metal by grinding or cutting. Mills are run on drill pipe or tubing to grind up debris in the hole, remove stuck portions of drill stem or sections of casing for sidetracking, and ream out tight spots in the casing. They are also called junk mills, reaming mills, and so forth, depending on what use they have. *v*: to use a mill to cut or grind metal objects that must be removed from a well.

milled bit *n*: also called a milled-tooth bit or a steel-tooth bit. See *steel-tooth bit*.

milled-tooth bit *n*: see *steel-tooth bit*.

modulus of elasticity *n*: the ratio between a force per unit area that deforms an object and the deformation caused by the force. Great force can be put on an object with a high modulus of elasticity, such as a diamond, before it is permanently deformed by the force.

mud *n*: the liquid circulated through the wellbore during rotary drilling and workover operations. In addition to its function of bringing cuttings to the surface, drilling mud cools and lubricates the bit and drill stem, protects against blowouts by holding back subsurface pressures, and deposits a mud cake on the wall of the borehole to prevent loss of fluids to the formation. Although it was originally a suspension of earth solids (especially clays) in water, the mud used in modern drilling operations is a more complex, three-phase mixture of liquids, reactive solids, and inert solids. The liquid phase may be fresh water, diesel oil, or crude oil and may contain one or more conditioners. See also *drilling fluid*.

N

nonreactive *adj*: as applied to drilling mud components, those components that are unaffected by and do not react with other components in the mud. Compare *reactive*.

normalizing *n*: heat-treating applied to metal tubular goods to ensure uniformity of the grain structure of the metal.

nose *n*: the pointed end of the cone of a roller cone bit or the rounded portion of the head of a diamond bit in which the diamonds are embedded.

nose button *n*: a hard-metal projection that is placed on the end of the pilot pin of a roller cone bit to absorb some of the wear created by outward thrusts as the bit rotates.

nozzle *n*: a passageway through jet bits that causes the drilling fluid to be ejected from the bit at high velocity. The jets of mud clear the bottom of the hole. Nozzles come in different sizes that can be interchanged on the bit to adjust the velocity with which the mud exits the bit.

O

off-center alignment *n*: see *cone alignment*.

off-center wear *n*: a type of bit wear in which the cutters on the cones wear in an uneven pattern because of the whirling action of the bit as it drills. Whirling is the motion a bit makes when it does not rotate around the center; instead it drills with a spiral motion.

offset roller cone bit *n*: a roller cone bit each cone of which displays offset from a center point when a line is extended through the middle of each cone. Offset roller cone bits are usually employed to drill soft formations, because the offset causes the teeth on the cone to gouge and scrape the formation. Gouging and scraping action is required to penetrate soft formations.

on-center alignment *n*: see *cone alignment*.

one-eyed jet bit *n*: see *jet deflection bit*.

operator *n*: the person or company who actually operates a well; generally, in countries where oil and gas operations are not a part of the government, an operator is a major (large) or independent (relatively small) oil company that leases the right to drill for and produce oil and gas from the individual or individuals who own the mineral rights to the land on which the well is located.

outer bearings *n pl*: the main bearings of a roller cone bit that reside inside the largest part of the cone. The outer bearings are either roller bearings or journal bearings. Compare *inner bearings*.

out-of-gauge bit *n*: a bit that is no longer of the proper diameter.

out-of-gauge hole *n*: a hole that is not in gauge—that is, it is smaller or larger than the diameter of the bit used to drill it.

overburden pressure *n*: the pressure exerted by the rock layers that overlie a formation of interest. It is generally considered to be about 1 pound per square inch per foot (22.62 kilopascals per metre). Note that overburden pressure is different from and independent of the subsurface pressure caused by fluids in a formation.

overgauge hole *n*: a hole whose diameter is larger than the diameter of the bit used to drill it. An overgauge hole can occur when a bit is not properly stabilized or does not have enough weight put on it.

P

packed-hole assembly *n*: a bottomhole assembly consisting of stabilizers and large-diameter drill collars arranged in a particular configuration to maintain drift angle and direction of a hole.

parabolic profile *n*: the shape of a PDC bit's head that resembles a parabolic arc, a curve that looks very much like an arch over a doorway. Compare *shallow-cone profile* and *short parabolic profile*.

PDC bit *n*: a special type of diamond drilling bit that does not use natural diamonds. Instead, polycrystalline diamond inserts (compacts) are embedded into a matrix on the bit. PDC bits are often used to drill very hard, abrasive formations, but also find use in drilling medium and soft formations.

penetration rate *n*: see *rate of penetration*.

permeability *n*: 1. a measure of the ease with which a fluid flows through the connecting pore spaces of rock. The unit of measurement is the millidarcy. 2. fluid conductivity of a porous medium. 3. ability of a fluid to flow within the interconnected pore network of a porous medium.

pin *n*: on a bit, the bit shank.

pin angle *n*: see *journal angle*.

pinched bit *n*: a roller cone bit on which the legs of the bit have been forced inward by a great force, such as that caused by jamming the bit into undergauge hole.

plasticity *n*: the ability of a substance to retain a shape that it has attained after being deformed. Compare *elasticity*.

polycrystalline diamond compact (PDC) *n*: a disk (a compact) of very small synthetic diamonds, metal powder, and tungsten carbide powder that are used as cutters on PDC bits. Compare *thermally stable polycrystalline diamond bit*.

polycrystalline diamonds *n pl*: very small synthetic diamonds formed by applying great heat and pressure to carbon.

porcupine bit *n*: name for a polycrystalline diamond compact (PDC) bit that has a particularly large number of PDC cutters installed on the head of the bit.

pore pressure *n*: the force exerted by the fluids in the pores of a formation, recorded in the wellbore at the depth of the formation with the well shut in. Also called formation pressure, reservoir pressure, or shut-in bottomhole pressure.

porosity *n*: 1. the condition of being porous (such as a rock formation). 2. the ratio of the volume of empty space to the volume of solid rock in a formation, indicating how much fluid a rock can hold.

positive-displacement downhole mud motor *n*: a device used to rotate the bit without rotating the drill stem. Basically, the motor comprises a spiral rod that is housed inside a helical-shaped chamber—the rod and chamber are a pump. Circulating mud down the drill stem and to the motor causes the rod to rotate. Since the bit is mechanically connected to the motor, as the motor rotates, so does the bit. This method of bit rotation eliminates the need to rotate the entire drill stem and thus is especially useful in directional drilling.

pressure loss *n*: a reduction in the amount of force a fluid exerts against a surface, such as the walls of a pipe. It usually occurs because the fluid is moving against the surface and is caused by the friction between the fluid and the surface.

profile *n*: see *shallow-cone profile, short parabolic profile, parabolic profile.*

R

race *n*: in a roller or ball bearing, the groove, or track, in which the bearings roll or rotate as the device in which they are installed operates.

radial flow *n*: in a diamond bit, the channels that direct the flow of drilling fluid across the cutting surface of the bit and in which the channels radiate out from the bit's center directly to the gauge area. Compare *feeder-collector.*

rake angle *n*: on polycrystalline diamond compact (PDC) bits, the angle at which the studs, on which are mounted the PDC cutters, are placed in the bit's head. If a stud has a rake angle of 90°, the stud is perpendicular to the plane of the bit's head. In most cases, the studs are mounted with back rake angle. See *back rake angle.*

rate of penetration (ROP) *n*: a measure of the speed at which the bit drills into formations, usually expressed in feet (metres) per hour or minutes per foot (metre).

reactive *adj*: as applied to drilling muds, those components in the mud that are affected by and react with other mud components.

ream *v*: to enlarge the wellbore by drilling it again with a special bit. Sometimes, an undergauge hole must be reamed.

reamer *n*: a tool used in drilling to smooth the wall of a well, enlarge the hole to the specified size, help stabilize the bit, straighten the wellbore if kinks or doglegs are encountered, and drill directionally. See *ream.*

reamer pad *n*: on a diamond bit, a flattened place above the bottomhole cutting surfaces whose purpose is to ream the hole above the bottom of the bit.

ridge set *n*: in diamond bits, the setting of small diamonds on raised ridges on the cutting surface of the bit. Compare *surface set.*

rock bit *n*: name for the first roller cone bits; now almost obsolete. See *roller cone bit.*

roller bearing *n*: a bearing in which the journal rotates in contact with a number of rollers usually contained in a cage. Compare *ball bearing, journal bearing.*

roller bit *n*: see *roller cone bit.*

roller cone bit *n*: a drilling bit made of two, three, or four cones, or cutters, that are mounted on extremely rugged bearings. The surface of each cone is made of rows of steel teeth or rows of tungsten carbide inserts. Also called rock bit.

roller race *n*: a track, channel, or groove in which roller bearings roll.

ROP *abbr*: rate of penetration.

S

sealed bearing *n*: on a roller cone bit, a type of bearing that is not exposed to the drilling fluid in the wellbore. Instead, a synthetic rubber ring is placed between the outside of each cone and the shirttail of the bit where the cones reside. This seal prevents drilling fluid from entering the space between the cone and shirttail and thus the bearings, which lie inside the cones. Sealed bearing bits must also contain a built-in reservoir of grease since they are not lubricated by the drilling fluid.

shallow-cone profile *n*: the shape of a PDC bit's head that resembles two short and rounded cones placed side by side. The compacts are placed on the surface of the head. Compare *parabolic profile*, *short parabolic profile*.

shear bit *n*: see *fixed-head bit*.

shirttail *n*: the part of a drilling bit on which the cone is anchored. Shirttails extend below the threaded pin of the bit and are usually rounded on bottom.

short parabolic profile *n*: the shape of a PDC bit's head in which the head resembles a short parabolic arc. A parabolic arc is an arch-shaped curve that looks similar to the arch over a doorway or other structure. Compare *parabolic profile*, *shallow-cone profile*.

shrouded jet nozzle *n*: a special type of jet nozzle that is manufactured with a projection (the shroud), which serves to minimize the erosion of the nozzle by the high-velocity jet of drilling fluid being forced through it.

side rake angle *n*: in a PDC bit, the left-to-right orientation of the cutter with respect to the bit's face; usually, the cutters are angled toward the outside of the bit to help direct cuttings to the annulus.

sidetracking bit *n*: a specially designed bit, usually made up on a downhole motor, that is used to drill a directional or horizontal well.

spall *n*: a condition in which the surface material of a bit bearing separates from the bearing's core material. Bearings spall when the metal of which they are made fatigues because of overuse.

spud bit *n*: a special kind of drilling bit with sharp blades rather than teeth. It is sometimes used for drilling soft, sticky formations.

steel-tooth bit *n*: a roller cone bit in which the surface of each cone is made up of rows of steel teeth. Also called a milled-tooth bit or milled bit.

stickiness *n*: the characteristic of a soft formation to adhere (stick) to the bit as the formation is drilled.

stiff drilling assembly *n*: a collection of downhole tools arranged in the drill stem above the bit to keep the hole on course. In general, a stiff drilling assembly consists of stabilizers and large-diameter drill collars. Also called packed-hole assembly.

stringer *n*: a relatively narrow section of a rock formation that interrupts the consistency of another formation and makes drilling that formation less predictable.

sub *n*: a short, threaded piece of pipe used to adapt parts of the drilling string that cannot otherwise be screwed together because of differences in thread size or design. A sub may also perform a special function, such as a bent sub, which deflects a downhole drilling motor off-vertical to begin drilling a directional hole.

surface set *n*: in diamond bits, the setting of relatively large diamonds into the cutting surface of the bit; the manufacturer buries about two-thirds of each diamond into the bit's matrix, leaving about one-third of each diamond exposed. Compare *ridge set*.

surfactant *n*: a soluble compound that concentrates on the surface boundary between two substances such as water and air and reduces the surface tension between the substances, allowing the two to thoroughly mix.

thermally stable polycrystalline (TSP) diamond bit *n*: a special type of fixed-head bit that has synthetic diamond cutters that do not disintegrate at high temperatures. Thermally stable diamonds are more like natural diamonds in that they are able to withstand relatively high temperatures without breaking apart. Compare *polycrystalline diamond compact bit*.

T

thermal stability *n*: a relative measure of the ability of a substance to withstand heat without disintegrating.

tight hole *n*: 1. a well about which information is restricted for security or competitive reasons. 2. a section of the hole that, for some reason, is undergauge. For example, a bit that is worn undergauge will drill a tight hole.

tool joint *n*: a heavy coupling element for drill pipe. Made of alloy steel, a tool joint has coarse, tapered threads and seating shoulders that sustain the weight of the drill stem, withstand the strain of frequent coupling and uncoupling, and provide a leakproof seal. The male section of the joint, or the pin, is attached to one end of a length of drill pipe, and the female section, or the box, is attached to the other end.

torque *n*: the turning force that is applied to a shaft or other rotary mechanism to cause it to rotate or tend to do so. Torque is measured in units of length and force (foot-pounds, newton-metres).

tracking *n*: a rare type of bit-tooth wear occurring when the pattern made on the bottom of the hole by all three cones of a rock bit matches the bit-tooth pattern to such an extent that the bit follows the groove or channel of the pattern and drills ahead very little.

tricone bit *n*: a type of roller cone bit in which three cone-shaped cutting devices are mounted in such a way that they intermesh and rotate together as the bit drills. The bit body may be fitted with nozzles, or jets, through which the drilling fluid is discharged.

trip *v*: to insert or remove the drill stem into or out of the hole.

true-to-gauge hole *n*: a hole that is the same size as the bit that was used to drill it. It is frequently referred to as a full-gauge hole.

TSP *abbr*: thermally stable polycrystalline. See *thermally stable polycrystalline diamond bit*.

tungsten carbide *n*: a fine, very hard, gray crystalline powder, a compound of tungsten and carbon. This compound is bonded with cobalt or nickel in cemented carbide compositions and used for cutting tools, abrasives, and dies.

tungsten carbide bit *n*: a type of roller cone bit with inserts made of tungsten carbide. Also called tungsten carbide insert bit.

tungsten carbide insert bit *n*: see *tungsten carbide bit*.

turbodrill *n*: a downhole motor that rotates a bit by the action of the drilling mud on turbine blades built into the tool. When a turbodrill is used, rotary motion is imparted only at the bit; therefore, it is unnecessary to rotate the drill stem. Although straight holes can be drilled with the tool, it is used most often in directional drilling.

two-cone bit *n*: a type of roller cone bit in which two cone-shaped cutting devices are mounted in such a way that they intermesh and rotate together as the bit drills. The bit body may be fitted with nozzles, or jets, through which the drilling fluid is discharged. Compare *tricone bit*.

UVW

undergauge bit *n*: a bit whose outside diameter is worn to the point at which it is smaller than it was when new. A hole drilled with an undergauge bit is said to be undergauge.

undergauge hole *n*: that portion of a borehole drilled with an undergauge bit.

velocity *n*: 1. speed. 2. the timed rate of linear motion.

wandering *n*: the tendency of the drill bit to deviate horizontally parallel to tilted strata.

washpipe *n*: in bits with watercourses instead of jet nozzles, a hardened erosion-resistant metal lining that is fitted inside the watercourses to prevent drilling mud from eroding the watercourses.

watercourse *n*: a hole inside a bit through which drilling fluid from the drill stem is directed.

wear pad *n*: a flat metal protrusion of a downhole tool that is usually hardfaced and that protects the rotating tool from wear.

weight on bit (WOB) *n*: the amount of downward force placed on the bit by the weight of the drill collars.

whirl *n*: see *bit whirl*.

wildcat well *n*: a well drilled in an area where no oil or gas production exists.

WOB *abbr*: weight on bit.

Review Questions
LESSONS IN ROTARY DRILLING
Unit I, Lesson 2: The Bit

With multiple-choice questions, place a mark in the blank space next to the phrase that best completes the sentence or answers the question. Note that more than one choice may be required. With true-or-false questions, place a T or an F in the appropriate blank space. With fill-in-the-blank questions, write the answers in the blank spaces provided.

1. To make a bit drill, the driller rotates it, puts weight on it, and—

    _____ a. adds tungsten carbide inserts.

    _____ b. uses diamonds as bit cutters.

    _____ c. forges teeth into the cones.

    _____ d. circulates drilling fluid out of it.

2. Undergauge hole is bad because—

    _____ a. the hole is too small to produce efficiently.

    _____ b. subsequent bits and other full-gauge tools lowered into it can get stuck.

    _____ c. it usually must be reamed, which wastes time.

    _____ d. both b and c

3. Rig operators have to change bits every time the wellbore encounters a different formation.

    True _____

    False _____

4. Cutter intermesh keeps—

    _____ a. the cutters sharp.

    _____ b. the cutters from breaking.

    _____ c. the cutters clean.

    _____ d. none of the above

5. Two types of roller cone bits are—

    a. _____

    b. _____

6. Tungsten carbide inserts are brittle and are therefore more likely to break under impact than steel teeth.

    True _____

    False _____

7. Bits for soft formations generally do not have much offset.

    True _____

    False _____

8. Put a check mark next to the words that describe the features of a steel-tooth bit that is designed to drill hard formations.

    _____ a. short teeth

    _____ b. long teeth

    _____ c. a lot of offset

    _____ d. little or no offset

    _____ e. tungsten carbide covering

9. Which tungsten carbide insert shape is best for drilling soft formations?

    _____ a. cone

    _____ b. chisel

    _____ c. hemispherical

    _____ d. none of the above

10. The cutters on the gauge row of a steel-tooth or insert bit do not intermesh.

    True _____

    False _____

11. Check the phrases that accurately describe the job of drilling fluid.

    _____ a. develops pressure to prevent kicks

    _____ b. goes into the formation to fill the pore spaces

    _____ c. carries cuttings away from the bit

    _____ d. coats the drill string to prevent erosion

    _____ e. cools the bit

12. Of the two ways drilling fluid exits the bit, which one usually cleans the hole better?

    _____ a. watercourses

    _____ b. jet nozzles

13. If you removed a new nozzle from its box and found the number 12 on it, you could assume that the nozzle's diameter is—

_____ a. $^{12}/_8$ of an inch (38.1 millimetres).

_____ b. $^{12}/_{16}$ of an inch (19.05 millimetres).

_____ c. $^{12}/_{32}$ of an inch (9.53 millimetres).

_____ d. $^{12}/_{64}$ of an inch (4.76 millimetres).

14. In general, decreasing a bit nozzle's size increases the velocity of the jet of drilling fluid leaving the nozzle.

True _____

False _____

15. In a journal bearing bit, what other type of bearing is sometimes present?

_____ a. roller

_____ b. ball

_____ c. line

_____ d. none of the above

16. A journal bearing is stronger than a roller bearing because it—

_____ a. has a large surface area.

_____ b. contacts the surface as a single line.

_____ c. is spherical in shape and contacts the surface as it rolls.

_____ d. is like rolling a glass across a table top.

17. Virtually all journal bearing bits are sealed.

True _____

False _____

18. The purpose of the pressure compensator in a sealed bearing bit is to—

_____ a. push the lubricant out of the grease reservoir and into the wellbore.

_____ b. keep the lubricant from getting to the bit's bearings.

_____ c. allow lubricant to flow from the grease reservoir to the bearings.

_____ d. allow lubricant to flow from the grease reservoir to the seal.

19. Bit whirl can sometimes be prevented by—

_____ a. reducing the weight on the bit and reducing the rpm.

_____ b. using a flexible bottomhole assembly.

_____ c. increasing the weight on the bit and increasing the rpm.

_____ d. using a stiff bottomhole assembly.

_____ e. Both a and b

_____ f. Both c and d

20. Broken inserts on a tungsten carbide bit are a normal wear characteristic in some formations.

    True _____

    False _____

21. Broken teeth on a steel-tooth bit are a normal wear characteristic in some formations.

    True _____

    False _____

22. Flat-crested wear and self-sharpening wear are normal on steel-tooth bits.

    True _____

    False _____

23. Name the three types of diamond bits.

    a. _____

    b. _____

    c. _____

24. Two properties of diamonds that make them wear out are their—

    _____ a. high impact strength and high thermal stability.

    _____ b. high impact strength and low thermal stability.

    _____ c. low impact strength and high thermal stability.

    _____ d. low impact strength and low thermal stability.

25. Which diamond-bit profile generally gives the longest life to the diamonds in the bit?

    _____ a. parabolic

    _____ b. single cone

    _____ c. double cone

    _____ d. concave

26. When drilling with a natural diamond bit in a soft formation, which diamond plot would probably work the best?

    _____ a. grid plot

    _____ b. circle plot

27. Of all the problems a PDC bit may encounter downhole, its worst enemy is—
    _____ a. hard, abrasive formation
    _____ b. soft, sticky shale
    _____ c. heat
    _____ d. pressure

28. Roller cone bit and PDC bit nozzles are interchangeable.
    True _____
    False _____

29. A TSP bit is as thermally stable as a natural diamond bit.
    True _____
    False _____

30. The two big enemies of diamond bits are—
    _____ a. spiral holes and nonmagnetic drill collars.
    _____ b. junk and heat.
    _____ c. cost and availability.
    _____ d. size and capability.

31. Eccentric bits drill undergauge hole to cure swelling problems.
    True _____
    False _____

32. Which one of the four types of bit drill a formation by shearing and slicing it?
    _____ a. TSP bits
    _____ b. roller cone bits
    _____ c. natural diamond bits
    _____ d. PDC bits

33. If a roller cone bit has a little less offset and slightly longer cutters than a soft-formation bit of the same type, the bit is probably designed to drill best in—
    _____ a. medium-soft formations.
    _____ b. medium-hard formations.
    _____ c. very soft formations.
    _____ d. extremely hard formations.

34. From an economic standpoint when drilling a hole, which is usually better?

    _____ a. A short-term improvement in ROP.

    _____ b. Good overall bit performance.

35. To determine the cost of drilling a foot (metre) of hole, contractors include—

    _____ a. cost of operating the rig, per hour

    _____ b. cost of the bit

    _____ c. trip time, in hours

    _____ d. drilling time, in hours

    _____ e. All of the above

—

Answers to Review Questions
LESSONS IN ROTARY DRILLING
Unit I, Lesson 2: The Bit

1. d
2. d
3. F
4. c
5. steel-tooth, or rock, bits; tungsten carbide insert, or button, bits
6. T
7. F
8. a, d, e
9. b
10. T
11. a, c, e
12. b
13. c
14. T
15. b
16. a
17. T
18. c
19. f
20. T
21. F
22. T
23. natural, polycrystalline diamond compact (PDC), thermally stable polycrystalline diamond (TSP)
24. d
25. b
26. a
27. c
28. F
29. F
30. b
31. F
32. d
33. a
34. b
35. e